A NOTE FROM THE AUTHORS

The AP Environmental Science course provides the principles, ideas, and processes required to understand the connections within the natural world. It also tries to give students the tools they need to identify natural and human-made problems, to analyze their causes, and to reflect upon alternative solutions for solving or averting them.

This book will help students take those skills and tools from the course and apply them to the AP exam, with focused summary and review. Success in any AP course requires diligence, discipline, and hard work, but the rewards can be great.

Best of luck,

Dora Barlaz

Craig C. Freudenrich

Jane P. Gardner

Kaplan offers resources and options to help you prepare for the PSAT, SAT, ACT, AP exams, and other high-stakes exams. Go to www.kaptest.com or scan this code below with your phone (you will need to download a QR code reader) for free events and promotions.

snap.vu/m87n

RELATED TITLES

AP Biology

AP Calculus AB & BC

AP Chemistry

AP English Language & Composition

AP English Literature & Composition

AP European History

AP Human Geography

AP Macroeconomics/Microeconomics

AP Physics B & C

AP Psychology

AP Statistics

AP U.S. Government & Politics

AP U.S. History

AP World History

SAT Premier with CD-Rom

SAT Strategies, Practice, and Review

SAT Subject Test: Biology E/M

SAT Subject Test: Chemistry

SAT Subject Test: Literature

SAT Subject Test: Mathematics Level 1

SAT Subject Test: Mathematics Level 2

SAT Subject Test: Physics

SAT Subject Test: Spanish

SAT Subject Test: U.S. History

SAT Subject Test: World History

AP® ENVIRONMENTAL SCIENCE

2014

Dora Barlaz

Craig C. Freudenrich

Jane P. Gardner

KAPLAN

PUBLISHING

New York

© 2013 by Kaplan, Inc.

Published by Kaplan Publishing, a division of Kaplan, Inc.
395 Hudson Street
New York, NY 10014

Printed in the United States of America

10 9 8 7 6 5 4 3 2 1

ISBN-13: 978-1-61865-254-6

Kaplan Publishing books are available at special quantity discounts to use for sales promotions, employee premiums, or educational purposes. For more informaton or to purchase books, please call the Simon & Schuster special sales department at 866-506-1949.

TABLE OF CONTENTS

PART THREE: AP ENVIRONMENTAL SCIENCE REVIEW

ABOUT THE AUTHORS

Dora Barlaz teaches at the Horace Mann School, an independent school in Riverdale, New York. She has taught AP Environmental Science since the inception of the course and has been an exam reader for four years. During the summer, she travels to exotic places.

Craig C. Freudenrich, PhD, earned a bachelor's degree in biology from West Virginia University and a doctorate in physiology from the University of Pittsburgh School of Medicine before completing eight years of postdoctoral research at Duke University Medical Center. He has conducted more than 20 years of biomedical research in neuroscience, plant physiology, muscle contraction, diabetes, and cell physiology. Craig also taught high school science in Durham, North Carolina, as well as the Science of Science Fiction at the Talent Identification Program at Duke University. He is interested in all sciences, especially biology, chemistry, physics, astronomy, and space exploration.

Jane P. Gardner works as a freelance science writer. She has a master's degree in geology from Virginia Tech and a master's degree in secondary education from the University of Massachusetts–Boston. She taught courses in geology, environmental science, Earth science, and biology at the high school and college levels for six years.

KAPLAN PANEL OF AP EXPERTS

Congratulations—you have chosen Kaplan to help you get a top score on your AP exam.

Kaplan understands your goals and what you're up against: achieving college credit and conquering a tough test while participating in everything else that high school has to offer.

You expect realistic practice, authoritative advice, and accurate, up-to-the-minute information on the test. And that's exactly what you'll find in this book, as well as every other book in the AP series. To help you (and us!) reach these goals, we have sought out leaders in the AP community and experts in environmental science. Allow us to introduce our experts:

AP ENVIRONMENTAL SCIENCE EXPERT

Dora Barlaz teaches at the Horace Mann School, an independent school in Riverdale, New York. She has taught AP Environmental Science since the inception of the course, and has been an AP reader for four years.

ENVIRONMENTAL SCIENCE EXPERT

James A. MacDonald, PhD, earned his bachelor's degree in environmental biology from Columbia University and a doctorate in ecology and evolution from Rutgers University. He has worked in conservation and natural resource management for more than 11 years, conducting ecological field research in five countries. James has extensive interests in the field of natural resource management, including fisheries management and regulation, invasive species, and environmental education. He has taught environmental science, genetics, ichthyology, ecology, and several other courses at the college level. He is currently working as a fisheries biologist in New York.

| Part One |

THE BASICS

CHAPTER 1: INSIDE THE AP ENVIRONMENTAL SCIENCE EXAM

INTRODUCTION

Congratulations on your decision to take the Advanced Placement (AP) Environmental Science exam. The test measures your knowledge on topics ranging from global climate change to population trends. After taking a college-level class, you certainly will have a large base of environmental science knowledge heading into the exam. However, the AP exam asks you to take things one step further—to apply that knowledge in complex and analytical situations to show evidence of college-level learning.

This guide offers not only a full review of AP Environmental Science but, more importantly, specific skills and strategies successful students use to score higher on the AP exam. In the following chapters, you will encounter reading strategies for your day-to-day assignments, writing strategies for the free-response questions (FRQs), and analytical skills for both FRQs and multiple-choice questions. Rote memorization of facts alone does not ensure success. While a solid foundation of environmental science knowledge is critical to your learning, application of that knowledge earns top scores. Keep the skills and strategies you learn in Chapter 2 of this guide in mind as you take the diagnostic quiz in Part Two to assess where you stand before you tackle the course review in Part Three.

Part Three is the meat of this guide. Its units and chapters are arranged chronologically to correlate with your classroom textbook and the College Board's course description guide for the AP exam. We have made sure to give you a solid review of the most-tested vocabulary, concepts, and issues on the AP exam. Each chapter contains review questions and detailed explanations of test items to help you determine the need for further study.

Finally, Part Four offers two full-length practice tests that closely mirror the actual AP exam, with test-like FRQs and detailed answer explanations. Unlike the biology, chemistry, and physics exams, which have existed for many years, there have been only two fully released AP Environmental Science exams.

Are you ready for your adventure through the study and mastery of everything AP Environmental Science? Good luck!

TEST PREP ≠ STUDYING

If you're holding this book, chances are you are already completing the AP Environmental Science course and gearing up for the AP Environmental Science exam. Your teacher has spent the year teaching you the material you will need to have at your disposal. But there is more to the AP Environmental Science exam than environmental know-how. You have to be able to work around the challenges and pitfalls of the test—and there are many—if you want your score to reflect your abilities. You see, studying environmental science is only part of the preparation for the AP Environmental Science exam. Rereading your textbook is helpful, but it's not enough.

That's where this book comes in. We'll show you how to use your knowledge of environmental science and put it to brilliant use on Test Day. We'll explain the ins and outs of the test structure and question format so you won't experience any nasty surprises. We'll even give you answering strategies designed specifically for the AP Environmental Science exam.

Preparing effectively for the AP Environmental Science exam means doing some extra work. You need to review your text and master the material in this book.

OVERVIEW OF THE TEST STRUCTURE

AP exams have been around for half a century. While the format and content have changed over the years, the basic goal of the AP program remains the same: to give high school students a chance to earn college credit or advanced placement. To do this, a student needs to do two things:

1. Find a college that accepts AP scores
2. Do well enough on the exam

The first part is easy because a majority of colleges accepts AP scores in some form or another. The second part requires a little more effort. If you have worked diligently all year in your course work, you've laid the groundwork. The next step is familiarizing yourself with the test.

WHAT'S ON THE TEST

The Educational Testing Service (ETS)—the company that creates the AP exams—releases a list of the topics covered on the exam. ETS even provides the percentage of the test questions drawn from each topic. Because this information is useful to anyone considering taking the test, review the following breakdown. The College Board is the organization that administers the Advanced Placement program. ETS is the company that generates the actual AP exams.

Topics Covered on the AP Environmental Science Exam

Earth Systems and Resources (10%–15%)

- Earth science
- Atmospheric science
- Water
- Soil

Living World (10%–15%)

- Ecosystems and biomes
- Energy flow
- Biodiversity
- Ecosystem change
- Biogeochemical cycles

Population (10%–15%)

- Population biology
- Human population issues

Use of Land and Water (10%–15%)

Energy (10%–15%)

- Fossil fuels
- Nuclear energy
- Renewable energy sources
- Conservation

Pollution (10%–15%)

- Air
- Water
- Soil
- Waste disposal
- Health impacts

Global Change (10%–15%)

- Global warming and ozone depletion
- Biodiversity loss
- Sustainable strategies and conservation
- Fishing
- Economics
- Urban sprawl
- Agriculture
- Logging, mining, grazing

As in all AP science subjects, laboratory work forms an essential part of the course. AP Environmental Science is unique in its inclusion of fieldwork investigations and trips to local institutions. The nature of this course component varies by region, but it is always linked to a major concept in the course. Out-of-school experiences may be as varied as a tour of a wastewater treatment plant, a workday on an organic farm, a visit to an animal sanctuary, a botanical survey of a forest, or an analysis of soil and water samples on the school field. All these experiences are designed to enhance AP Environmental Science in ways that are unavailable in the indoor classroom.

In addition to factual knowledge, the College Board goals are to identify and analyze environmental problems and to understand solutions. Six themes are designed to promote these goals:

1. Science as a process
2. Energy flow
3. Interconnectedness of all aspects of Earth
4. Consequences of human actions on nature
5. Cultural context of environmental factors
6. Sustainability as a necessity for survival

The chapters in the review section of this book are designed to take advantage of this design by focusing on concepts and synthesizing information from different concepts to better understand, and learn, the AP Environmental Science course and exam content.

Now that you know what's on the test, let's talk about the test itself.

The AP Environmental Science exam is three hours long. Half of the time is devoted to multiple-choice questions and half to free-response questions. The multiple-choice section consists of 100 questions and accounts for 60 percent of the score. The number of questions on each topic reflects the percentages listed under Topics Covered on pages 5–6.

Free-response questions stress the application of the principles of AP Environmental Science and require the correlation of information from several areas. For example, a question about the development of a pristine area may involve knowledge of ecosystems, susceptibility of species to endangerment, and legislation relevant to the case.

There are four free-response questions: one document-based, one data set, and two synthesis-and-evaluation questions. Reasoning and analytical skills, as well as actual familiarity with the subject matter, are being measured. Calculators are not allowed. This section counts for 40 percent of the score.

Free-response questions allow for many correct answers. For example, when asked for a law, a place, an animal, or a plant, there can be many correct answers. Although the majority of students respond with a certain range of answers, it is understood that because students are taking these tests around the world and may be exposed to different examples of these concepts, the reading consultants will research obscure answers.

The term "free response" means roughly the same thing as "large, multistep, and involved," because you will spend the 90 minutes of Section II answering these four questions. Although these free-response questions are long and often broken down into multiple parts, they usually don't cover an obscure topic. Instead, they take a fairly basic concept and ask you a *bunch* of questions about it. Sometimes diagrams are required, or experiments must be outlined properly.

HOW THE EXAM IS SCORED

Scores are based on the number of questions answered correctly. **No points are deducted for wrong answers.** No points are awarded for unanswered questions. Therefore, you should answer every question, even if you have to guess.

When your three hours of testing are up, your exam is sent away for grading. The multiple-choice part is handled by a machine, while qualified graders—a group that includes teachers and professors, both current and former—grade your responses to Section II. After an interminable wait, your composite score will arrive by mail. (For information on rush score reports and other grading options, visit collegeboard.com or ask your AP Coordinator.) Your results will be placed into one of the following categories, reported on a five-point scale:

5 = Extremely well qualified (to receive college credit or advanced placement)

4 = Well qualified

3 = Qualified

2 = Possibly qualified

1 = No recommendation

Some colleges will give you credit for a score of 3 or higher, but it's much safer to get a 4 or a 5. If you have an idea of where you will be applying to college, check out the school's website or call the admissions office to find out their particular rules regarding AP scores. For information on scoring the practice tests in this book, see "How to Compute Your Score" in Part Four, page 293.

REGISTRATION AND FEES

You can register for the AP Environmental Science exam by contacting your guidance counselor or AP Coordinator. If your school doesn't administer the exam, contact AP Services for a listing of schools in your area that do. The fee for each AP exam is $89, and $117 at schools and testing centers outside the United States. For those qualified with acute financial need, the College Board offers a $28 credit. In addition, most states offer exam subsidies to cover all or part of the remaining cost for eligible students. To learn about other sources of financial aid, contact your AP Coordinator.

For more information on all things AP, visit collegeboard.com or contact AP Services:

AP Services
P.O. Box 6671
Princeton, NJ 08541-6671
Phone: 1-609-771-7300 or 1-888-225-5427 (toll-free in the United States and Canada)
email: apexams@info.collegeboard.org

WHAT TO BRING

Testing conditions vary from site to site. However, there are several key items that all students should bring on Test Day:

- Several sharpened no. 2 pencils with erasers. We suggest that you also bring along a separate eraser. White erasers work very well in erasing pencil from scan sheets.
- Several black or dark blue ballpoint ink pens. We suggest that you not use erasable ink or liquid ink pens. They can smear or run, affecting legibility.
- Your school code
- A watch that does not make noise. Some testing sites do not have clocks visible.
- Your Social Security number for identification purposes
- A photo ID

There are items that are prohibited or best left at home. These are items not to bring:

- Books, correcting fluid, dictionaries, highlighters, or notes

- Scratch paper

- Computers (unless you are a student with a disability and have been approved to bring the computer)

- Watches that beep or have an alarm

- Portable listening devices, such as MP3 players, CD players, radios, and tape players

- Cameras

- Beepers, cellular phones, or personal digital assistants (PDAs)

- Clothing with subject-related information

ADDITIONAL RESOURCES

The College Board website at www.collegeboard.com/ap/student/testing/ap/about.html is the best resource for additional information regarding AP courses, exams, and services. We suggest you visit often throughout the school year to access information regarding updates and test dates, and to answer any questions you may have along the way.

CHAPTER 2: STRATEGIES FOR SUCCESS: IT'S NOT ALWAYS HOW MUCH YOU KNOW

Changes that you noticed in your own water and air may be what led some of you to register for AP Environmental Science. Now, armed with the information you have learned in the course, you will have the opportunity to demonstrate what you have learned.

Remember, your learning goes far beyond the taking of this test. You will discover soon, if you have not already, that your acquaintances will look to you for explanations and analyses of global environmental change. In addition, as an informed citizen, you will be making responsible choices and decisions in your own life.

Because all of the College Board–approved AP Environmental Science textbooks cover the required information, you can start your studying in the source with which you are most familiar. A quick review of all subjects covered should lead you easily into AP Environmental Science.

GENERAL TEST-TAKING STRATEGIES

1. **Pacing**. Because many tests are timed, proper pacing allows the test taker to attempt every question in the time allotted. Poor pacing causes students to spend too much time on some questions to the point where they run out of time before completing all the questions.

2. **Process of Elimination**. On every multiple-choice test you take, the answer is given to you. The only difficulty resides in the fact that the correct answer is hidden among incorrect choices. Even so, the multiple-choice format means you don't have to pluck the answer out of the air. Instead, you can eliminate answer choices you know are incorrect, and when only one choice remains, then that must be the correct answer.

3. **Knowing When to Guess**. The AP Environmental Science exam does not deduct points for wrong answers, while questions left unanswered receive zero points. That means you

should always guess on a question you can't answer any other way. Over time, if you practice educated guessing, you'll see your score rise.

4. **Patterns and Trends**. The key word here is the *standardized* in "standardized testing." Being standardized means that tests don't change greatly from year to year. Sure, each question won't be the same, and different topics will be covered from one administration to the next, but there will also be a lot of overlap from one year to the next. That's the nature of *standardized* testing: If the test changed wildly each time it came out, it would be useless as a tool for comparison. Because of this, certain patterns can be uncovered regarding any standardized test. Learning about these trends and patterns can help if you are taking the test for the first time.

5. **The Right Approach**. Having the right mind-set plays a large part in how well you do on a test. If you are nervous about the exam and hesitant to make guesses, you may fare much worse than if you have an aggressive, confident attitude. Students who start with the first question and plod on from there don't score as well as students who deal with the easy questions before tackling the harder ones. People who take a test cold have more problems than those who take the time to learn about the test beforehand. In the end, factors like these make the difference between people who are good test takers and those who struggle even when they know the material.

These points are all valid for every standardized test, but they are quite broad in scope. The rest of this chapter will discuss how these general ideas can be modified to apply specifically to the AP Environmental Science exam. These test-specific strategies and the factual information covered in your course, as well as this book's review section, are the one-two punch that will help you succeed on the exam.

HOW TO APPROACH THE MULTIPLE-CHOICE QUESTIONS

Because all environmental scientists use "the scientific method" in some form or another, let's apply it to the test. Our hypothesis is that you can achieve a higher score by attacking the 100 multiple-choice questions in a specific order. Other students will just turn the page once the test begins and start with question 1. Let these people be the control group.

All 100 questions are multiple-choice questions, but there are three distinct question types:

1. **Stand-Alone Questions**. These are the first questions on each AP Environmental Science exam, and they typically make up a little over half of the exam. Each Stand-Alone covers a specific topic, and then the next Stand-Alone hits a different topic. Usually there are 5 to 60 words in the question stem, and these words provide you with the information you need to answer the question. Here is a typical Stand-Alone:

44. Which of the following is a carbon sink?

  (A) Volcanoes
  (B) All sedimentary rocks
  (C) Windmills
  (D) Trees
  (E) Most igneous rocks

The number of the question, 44, makes no difference because there's no order of difficulty on the AP Environmental Science exam. Tough questions are scattered between easy and medium questions.

2. **Cluster Questions**. With these questions, you get some initial piece of information (often visual), and this information has the letters (A) through (E) within it. Four to five problems without answer choices follow this information, and each question describes one of the choices (A–E) in the initial information. You pick the right letter choice for each problem. Here is an example:

  (A) Dennis Hayes
  (B) Rachel Carson
  (C) Aldo Leopold
  (D) Theodore Roosevelt
  (E) Henry David Thoreau

56. Established the first Earth Day, 1970

57. Wrote a *Sand Country Almanac* about ecological restoration

58. Promoted a back-to-nature attitude in the 19th century

59. President in the early 1900s who promoted establishment of national parks

This shows a typical group of Cluster Questions. The initial information lists five people as answer choices, and then the four problems ask you to pick the right people.

3. **Data Questions**. Data Questions appear on the AP Environmental Science exam. Just as the name suggests, a group of two to four questions is preceded by data in one form or another. The data might be a simple sentence or two, but usually it is something more complex, such as:

• A description of an experiment (50–200 words), often with an accompanying illustration

• A graph or series of graphs

• A large table

• A diagram

The next question is a sample Data Question.

Questions 98–99 refer to the following population pyramids showing age-sex distributions.

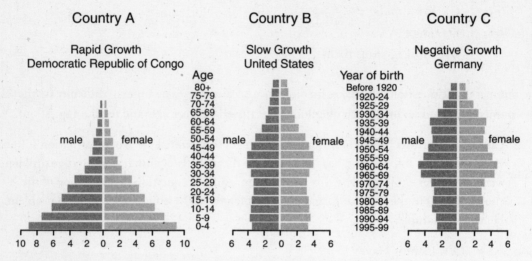

Source: United Nations, *World Population Prospects, The 1998 Revision.*

98. According to the diagrams above, what can be said about the population of country A?

   (A) Country A has a declining population.

   (B) Country A has a high death rate among children under the age of 5.

   (C) Citizens in country A have a greater than average annual income.

   (D) Citizens in country A have a long life expectancy.

   (E) Citizens in country A have very healthy diets.

99. Country C is characterized by which of the following?

   (A) Short life expectancy

   (B) 50 percent illiteracy rate

   (C) Few opportunities for women

   (D) Declining population

   (E) High emigration

You might think there is an advantage in the fact that the questions all concern the same set of data. However, Data Questions require a student to analyze the presented information carefully, and while that is occurring, time is ticking away and no questions are being answered.

Two common strategies may be used. Each student must choose the one that is best suited to his or her own learning style. If you prefer to use the standard strategy that you have been taught to use on tests throughout your school years, that strategy is fine. Answer the questions you know immediately, which probably should include the questions that involve simple reading of data, then go back to answer the more complicated ones. Finally, try to answer the more time-consuming questions that require much more logic and the process of elimination.

Those are the three question types. Combine this knowledge with the fact that you must answer 100 questions in 90 minutes, and you'll see why this game plan makes sense:

1. Always start with the Cluster Questions, because they take the least amount of time.

2. Then tackle the Stand-Alones.

3. Finally, use the time remaining to attempt some (but not necessarily all) of the Data Questions.

You don't need to get every multiple-choice question right to score well on the AP Environmental Science exam. A grade of 4 or 5 requires that you get a large portion, but not all, of the multiple-choice questions right. If you don't have enough time to get to every question, make sure that the questions you skip are the longest, most involved ones (the Data Questions). If you understand the topic, you can answer five Cluster Questions in roughly a minute. That's a great use of your limited resource: time.

So while the control-group students around you are blithely starting with question 1 and proceeding from there, you are going to head to the Cluster Questions and answer them first. Once you're done with those easy pickings, it's off to the Stand-Alones, and finally the Data Questions. This approach will make the best use of your time and effort on the AP Environmental Science exam.

There's more to it than just tackling questions in the right order, however. The more you know about each question type, the better equipped you will be to handle it.

CLUSTER QUESTIONS

You have two main options on how you should approach a group of short, sweet Cluster Questions. If you look at the cluster and know the topic well, start with the answer choices (A–E) and then find the question that matches each one. For instance, if you are confident you know who each of these five people are . . .

(A) Dennis Hayes
(B) Rachel Carson
(C) Aldo Leopold
(D) Theodore Roosevelt
(E) Henry David Thoreau

. . . then start with (A), Hayes, and seek out a question that matches it. You'll find question 56. One question down, three to go. The advantage to this approach is that it puts you in control of the test. The more you break away from that control group "answer every question the way everybody else does" mind-set, the better your score will be.

Sometimes the answers-then-question approach isn't the best. This is especially true if you don't know the topic or answer choices as well. Let's say you know who (A) and (E) are, but you're vague about the other three people. This means you get the answers to 56 (Hayes) and 58 (Thoreau) correct, but you don't know the answers to 57 and 59. Should you leave them blank? No! For both questions, you can eliminate the people you know are incorrect [(A) and (C)], so you should take a guess and hope for the best.

Anytime you can eliminate at least one answer choice from a problem, you should take a guess. You won't get every guess right, but over the course of the test this form of educated guessing will improve your score. Let's say you picked (B) for 57 and (D) for 59. It turns out the answer for 57 is (C) and 59 is (D), so you got one right and one wrong. Overall, that boosts your score.

The same letter can be the correct answer for more than one problem in the same cluster. It doesn't happen very often, but it does occur, so don't worry if you find yourself picking the same letter twice.

STAND-ALONE QUESTIONS

It's easier to talk about what isn't in the Stand-Alone Questions than what is there:

- There's no order of difficulty; that is, questions don't start out easy and gradually become tougher.
- There are no two questions connected to each other in any way.
- There's no pattern as to what environmental science concepts appear when.

The Stand-Alones look like a bunch of disconnected environmental science questions, one following another, and that's just what they are. A question about the atmosphere may follow a question about water, which may follow a question about endangered species.

There's no overall pattern. But just because the section is randomly ordered, that doesn't mean you have to approach it on the same random terms. Instead, draw up two lists right now using the topics covered in Chapter 1. Label one list "Concepts I Enjoy and Know About Environmental Science" and label the other list "Concepts That Are Not My Strong Points."

When you get ready to tackle the Stand-Alone section, keep these two lists in mind. On your first pass through the section, answer all the questions that deal with concepts you like and know a lot about. None of the questions will have a lot of text in front of it, so you should be able to figure out very quickly whether you have the facts needed to answer it. If you do, answer that problem and move on. If the question is on a subject that's not one of your strong points, skip it and come back later.

The overarching goal is to answer correctly the greatest possible number of questions in the time available. To do this, focus on your strengths during the first pass through the Stand-Alones. Some questions might be very difficult, even in a subject you're familiar with. Take a minute or so on a tough question, and if you can't come up with an answer, make a mark by the question number in your test booklet and move on. The first pass is about picking up easy points.

Once you've swept through and snagged all the easy questions, take a second pass and try the tougher ones. These tougher questions might cover subjects you're not strong in, or they might just be very difficult questions on subjects you are familiar with. Odds are high that you won't know the answer to some of these questions, but don't leave them blank. You should always take a stab at eliminating some answer choices, and then make an educated guess.

Admittedly, the AP Environmental Science exam is a test of specific knowledge, so picking the right answer from the incorrect answer choices is harder to do than it is on other standardized tests. Still, it can be done. The following two key ideas are easy to remember and will come in handy on the tougher multiple-choice questions.

COMPREHENSIVE, NOT SNEAKY

Some tests are sneakier than others. They have convoluted writing, questions designed to trip you up mentally, and a host of other little tricks. Students taking a sneaky test often have the proper facts, but they get the answer wrong because of a trap in the question itself.

The AP Environmental Science exam is *not* a sneaky test. Its objective is to see how much environmental science knowledge you have stored in that brain of yours. To do this, it asks a wide range of questions from an even wider range of environmental science topics. The exam tries to cover as many different facts in environmental science as it can, which is why the problems jump around. The test works hard to be as comprehensive as it can be, so that students who only know one or two environmental science topics will soon find themselves struggling.

Understanding these facts about how the test is designed can help you answer questions on it. The AP Environmental Science exam is comprehensive, not sneaky; it makes questions hard by asking about hard subjects, not by crafting hard questions. And you've probably taken an AP Environmental Science course, so trust your instincts when guessing. If you think you know the right answer, chances are you dimly remember the topic being discussed in your AP course. The test is about knowledge, not traps, so trusting your instincts will help more often than not.

You don't have much time to ponder every tough question, so trusting your instincts can help keep you from getting bogged down and wasting time on a problem. You might not get every educated guess correct, but again, the point isn't about getting a perfect score. It's about getting a good score, and surviving hard questions by going with your gut feelings is a good way to achieve this.

On other problems, though, you might have no inkling of what the correct answer should be. In that case, turn to the second key idea.

THINK "GOOD SCIENCE!"

Good common science sense as well as information you have learned in previous science classes should always be used.

The AP Environmental Science exam rewards good scientists. The exam wants to foster future environmental scientists by covering fundamental topics and sound laboratory procedure. What the exam doesn't want is bad science. It doesn't want answers that are factually incorrect, too extreme to be true, or irrelevant to the topic at hand.

Yet these "bad science" answers invariably appear, because it's a multiple-choice exam and you have to have four incorrect answer choices around the one right answer. So if you don't know how to answer a question, look at the answer choices and think "good science." This may lead you to find some poor answer choices that can be eliminated. Here is an example:

22. Ethanol is

   (A) a liquid produced by fermentation of plant matter.
   (B) used to generate most of the electricity used in California.
   (C) a fossil fuel.
   (D) a form of nuclear energy.
   (E) more commonly used than petroleum.

Even if you are not quite sure what ethanol is, you have probably heard of it in the news about energy alternatives, so that should eliminate (C). You know that gasoline refined from petroleum is what you put into your car at the pump, so that eliminates (E). It is true that the use of ethanol is on the rise as the price of petroleum rises, but even California, which has progressive air-quality laws, has not converted heavily, so that eliminates (B). The chemical suffix indicates that the word is some type of alcohol, which implies fermentation of plant matter. This is the "good science" answer.

You would be surprised how many times the correct answer on a multiple-choice question is a simple, blandly worded fact.

Thinking about "good science" in terms of the AP Environmental Science exam can help you in two ways:

1. It helps you cross out extreme answer choices or choices that are untrue or out of place.

2. It can occasionally point you toward the correct answer because the correct answer will be a factual piece of information sensibly worded.

Neither the "good science" nor the "comprehensive, not sneaky" strategy is 100 percent effective every time, but they do help more often than not. On a tough Stand-Alone Question, these techniques can make the difference between a random guess and a good guess.

QUESTIONS WITH DATA, GRAPHS, DESIGNING EXPERIMENTS

Data Questions require the most time and effort, but they aren't worth more than other multiple-choice question types. This is why you should save them for last. When you finish the Stand-Alone Questions, ideally you should have at least one minute for every Data Question remaining. You might have more time, which would be good; if you have less, you'll have to make some choices. Either way, scan the section and look for the topics you are most comfortable with. If you have to deal with a diagram, try to eliminate the most obvious answers first, then try to eliminate each of the harder answers one by one. Diagrams are usually very straightforward. Headings on charts, labels on axes of graphs, and units are all critical parts of a question or answer.

At least one of the Data Questions will usually deal with an experiment. Make sure you understand all the basic points of an experiment—testing a hypothesis, setting up an experiment properly to isolate a particular variable, and the meaning of controls, constraints, and conclusions—so that you will be able to tackle this section successfully. The key to getting through the Data Questions section of the exam is to be able to quickly analyze and draw a conclusion from data presented.

QUESTIONS WITH GRAPHS

There aren't many graphs on the AP Environmental Science exam, but when they appear, they are usually in the Data Questions section. Most graph questions usually require a bit of knowledge to determine what the right answer is, but some graph questions only test whether you can read a graph properly. If you can make sense of the vertical and horizontal axes, then you can determine what the correct answer is. Granted, very few graph questions are this easy, but even so, it's nice to have a slam dunk question or two. Therefore, if you see a graph, look at the problem and see if you can answer the question just by knowing how to read a graph.

That's all that can be said for the multiple-choice section of the AP Environmental Science exam. Be sure to practice these strategies on the practice test in this book so that you'll actually use them on the real test. Once you implement these techniques, your mind-set, approach, and score should benefit.

Of course, the multiple-choice questions only account for 60 percent of your total score. To get the other 40 percent, you have to tackle the free-response questions.

HOW TO APPROACH THE 10-MINUTE READING PERIOD

Some people like to take a few minutes to plan answers. Read and reread each question. Make sure you understand what is being asked. Next you should jot down any thoughts you have about the answer on a piece of scratch paper. Write down key words you want to mention. This is exactly what you want to do here as well: Brainstorm ideas about the best way to answer each free-response question.

> Waste disposal in the United States is improving. Today, more than half of the solid U.S. waste produced is brought to sanitary landfills. There are a number of structural and protective features incorporated into a well-designed sanitary landfill to alleviate problems that existed in the past.
>
> (A) (3 points) Describe three problems associated with all landfills.
> (B) (3 points) For each of these problems, connect it with the structure in the sanitary landfills that can alleviate or minimize the problem.
> (C) (3 points) Name three ways to reduce the amount of waste generated.
> (D) (2 points) Name and describe one law or treaty that deals with solid waste.
>
> Notes:
> (A) smelly, rats, methane, leakage
> (B) liners, pipes
> (C) reduce, reuse, recycle, dematerialization
> (D) Superfund (specifically toxic waste), Pollution Prevention Act of 1990

By the time you've written that, start writing the answer. When the time comes to start writing your answers, you'll have a good set of notes on which to base your answer to this question.

HOW TO APPROACH THE FREE-RESPONSE QUESTIONS (FRQ)

For Part Two, you have 90 minutes to jot down answers to only four questions. That's a little over 20 minutes per question, which gives you an idea of how much work each question may take. You might finish faster, but it wouldn't be good to finish each free-response question in less than 5 minutes because each question counts as 10 percent of your final score. Take the time to make your answers as precise and detailed as possible.

The College Board states that the four free-response questions consist of the following formats: one data-set question, one document-based question, and two synthesis-and-evaluation questions.

The data-set and document-based questions are platforms for you to demonstrate that you can effectively analyze data in various formats and from diverse sources (numerals and prose). The synthesis-and-evaluation questions require that you interpret and synthesize material from several sources to create convincing, logical essays. Free-response questions can come in any shape, size, or topic; however, there are details about this particular type of question that will be helpful to know beforehand.

IMPORTANT DISTINCTIONS

Each free-response question will, of course, be about a distinct topic. However, this is not the only way in which these questions differ from one another. Each question will also need a certain kind of answer, depending on the type of question it is. Part of answering each question correctly is understanding what general type of answer is required. Here are five important signal words that indicate the rough shape of the answer you should provide:

1. Describe
2. Discuss
3. Explain
4. Compare
5. Contrast

Each of these words indicates that a specific sort of response is required; none of them mean the same thing. Questions that ask you to *describe, discuss,* or *explain* are testing your comprehension of a topic. A *description* is a detailed verbal picture of something; a description question is generally asking for "just the facts." This is not the place for opinions or speculation. Instead, you want to create a precise picture of something's features and qualities. A description question might, for example, ask you to describe the results you would expect from an experiment. A good answer here will provide a rich, detailed account of the results you anticipate.

A question that asks you to *discuss* a topic is asking you for something broader than a mere description. A discussion is more like a conversation among ideas, and—depending on the topic—this may be an appropriate place to talk about tension between competing theories and views. For example, a discussion question might ask you to discuss whether it is best to recycle paper or produce new paper products from trees. A good answer here would go into detail about why one does less harm to the environment than the other.

A question that asks you to *explain* something is asking you to take something complicated or unclear and present it in simpler terms. For example, an explanation question might ask you to explain the reason why certain animals are more likely to become extinct. A simple listing of factors would not be an adequate answer to the question. Instead, you would need to state the factors *and* talk about why these factors put the species at a disadvantage.

> **AP EXPERT TIP**
>
> Prepare for each free-response question by underlining these action words. It's crucial that you answer the question asked—and by underlining these action words, you'll stay on the right track.

Questions that ask you to *compare* or *contrast* are asking you to analyze a topic in relation to something else. A question about *comparison* needs an answer that is focused on similarities between the two things. On the other hand, a question that focuses on *contrast* needs an answer emphasizing differences and distinctions.

TWO POINTS TO REMEMBER ABOUT FREE-RESPONSE QUESTIONS

1. **Most Questions Are Stuffed with Smaller Questions.** The typical question has an initial setup followed by questions labeled (A), (B), (C), and so on.

2. **Writing Smart Things Earns You Points!** For each sub-question on a free-response question, points are given for saying the right thing. The more points you score, the better off you are on that question. Going into the details about how points are scored would make your head spin, but in general, the AP Environmental Science people have a rubric that acts as a blueprint for what a good answer should look like. Every subsection of a question has two to five key ideas attached to it. If you write about one of those ideas, you earn yourself a point. There's a limit to how many points you can earn on a single sub-question, but it boils down to this: Writing smart things about each question will earn you points toward that question.

So don't be tense or in a hurry. You have 20 minutes to answer each free-response question. Use the time to be as precise as you can be for each sub-question. Part of being precise is presenting your answer in complete sentences. Do not simply make lists or outlines unless asked to do so. Sometimes doing well on one sub-question will earn you enough points to cover up for another sub-question you're not as strong on. When all the points are tallied for that free-response question, you come out strong on total points, even though you didn't ace every single sub-question.

There are questions in which the total number of points possible is greater than the 10 needed for a perfect score, but each subsection still has a maximum number of points. So don't spend all of your time on one subsection at the expense of another. Be sure to address everything being asked if you can. Always reach into your more creative thinking for some kind of response even if you don't think you know the answer.

Beyond these points, there's a bit of a risk in the free-response section because there are only four questions. If you get a question on a subject you're weak in, things might look grim. Still, take heart. Quite often, you'll earn some points on every question because there will be some sub-questions or segments that you are familiar with. Remember, the goal is not perfection. If you can ace two of the questions and slug your way to partial credit on the other two, you will put yourself in position to get a good score on the entire test. That's the Big Picture, so don't lose sight of it just because you don't know the answer to one sub-question.

AP EXPERT TIP

Unlike with multiple-choice, you can be more creative on the free-response questions if you're not sure about the answer. Just be sure to stick to the topic, and try to connect the question to something you remember about the topic.

Be sure to use all the strategies discussed in this chapter when taking the practice exams. Trying out the strategies there will get you comfortable with them, and you should be able to put them to good use on the real exam.

STRESS MANAGEMENT

You can beat anxiety to some degree—by being familiar with the subject matter and by developing strategies to deal with it.

SOURCES OF STRESS

In the space provided, write down your sources of test-related stress. The idea is to pin down any sources of anxiety so you can deal with them one by one. Following are common examples—feel free to use them and any others you think of:

- I always freeze up on tests.
- I'm nervous about the questions that involve chemistry.
- I need a good/great score to get into my first-choice college.
- My older brother/sister/best friend/girlfriend/boyfriend did really well. I must match that score or do better.
- My parents, who are paying for school, will be quite disappointed if I don't do well.
- I'm afraid of losing my focus and concentration.
- I'm afraid I'm not spending enough time preparing.
- I study like crazy but nothing seems to stick in my mind.
- I always run out of time and get panicky.
- The simple act of thinking, for me, is like wading through refrigerated honey.

MY SOURCES OF STRESS

_____

_____

_____

_____

_____

_____

_____

Read through your list. Cross out things or add things. Now rewrite the list in order of most disturbing to least disturbing.

MY SOURCES OF STRESS, IN ORDER

_____

_____

_____

_____

_____

_____

_____

Chances are, the top of the list is a fairly accurate description of exactly how you react to test anxiety, both physically and mentally. The later items usually describe your fears (disappointing Mom and Dad, looking bad, etc.). Taking care of the major items from the top of the list should go a long way toward relieving overall test anxiety. That's what we'll do next.

STRENGTHS AND WEAKNESSES

Take 60 seconds to list the areas of environmental science that you are good at. They can be general ("chemistry") or specific ("nitrogen cycle"). Put down as many as you can think of, and if possible, time yourself. Write for the entire time; don't stop writing until you've reached the one-minute stopping point. Go.

STRONG TEST SUBJECTS

_____

_____

_____

_____

_____

_____

_____

Now take one minute to list areas of the test you're not so good at, just plain bad at, have failed at, or keep failing at. Again, keep it to one minute, and continue writing until you reach the cutoff. Go.

WEAK TEST SUBJECTS

_____

_____

_____

_____

_____

_____

Taking stock of your assets and liabilities lets you know the areas you don't have to worry about, and the ones that will demand extra attention and effort. It helps a lot to find out where you need to spend extra effort. We mostly fear what we don't know and are probably afraid to face. You can't help feeling more confident when you know you're actively strengthening your chances of earning a higher overall score.

Now, go back to the "good" list, and expand on it for two minutes. Take the general items on that first list and make them more specific; take the specific items and expand them into more general conclusions. Naturally, if anything new comes to mind, jot it down. Focus all of your attention and effort on your strengths. Don't underestimate yourself or your abilities. Give yourself full credit. At the same time, don't list strengths you don't really have; you'll only be fooling yourself.

Expanding from general to specific might go as follows. If you listed "ecology" as a broad topic you feel strong in, you would then narrow your focus to include areas of this subject about which you are particularly knowledgeable. Your areas of strength might include population analysis, energy flow in communities, etc. Whatever you know well goes on your "good" list. OK. Check your starting time. Go.

STRONG TEST SUBJECTS: AN EXPANDED LIST

_____

_____

_____

_____

_____

_____

_____

_____

After you've stopped, check your time. Did you find yourself going beyond the two minutes allotted? Did you write down more things than you thought you knew? Is it possible you know more than you've given yourself credit for? Could that mean you've found a number of areas in which you feel strong?

You just took an active step toward helping yourself. Enjoy your increased feelings of confidence, and use them when you take the AP Environmental Science exam.

COUNTDOWN TO THE TEST

STUDY SCHEDULE

The schedule presented here is the ideal. Compress the schedule to fit your needs. Do keep in mind, though, that research in cognitive psychology has shown that the best way to acquire a great deal of information about a topic is to prepare over a long period of time. Because you may have several months to prepare for this exam, it makes sense for you to use that time to your advantage. This book, along with your text, should be invaluable in helping you prepare for this test.

If you have two semesters to prepare, use the following schedule:

September:
Take the diagnostic test in this book and isolate areas in which you need help. The diagnostic will serve to familiarize you with the type of material you will be asked about on the AP exam. Begin reading your environmental science textbook along with the class outline.

October–February:
Continue reading this book and use the summaries at the end of each chapter to help guide you to the most salient information for the exam.

March and April:
Take the two practice tests and get an idea of your score. Also, identify the areas in which you need to brush up. Then go back and review those topics in both this book and your environmental science textbook.

AP EXPERT TIP

In your AP class, pay close attention to your tests. For answers you get wrong, make sure you look up the correct answers yourself, rather than asking your teacher. This will help you remember that same information when it comes time for the AP exam.

May:

Do a final review and take the exam.

If you only have one semester to prepare, you'll need a more compact schedule:

January:

Take the diagnostic test in this book.

February–April:

Begin reading this book and identify areas of strength and weakness.

Late April:

Take the two practice tests and use your performance results to guide you in your preparation.

May:

Do a final review and take the exam.

THREE DAYS BEFORE THE TEST

It's almost over. Eat a PowerBar, drink some soda—do whatever it takes to keep going. Here are Kaplan's strategies for the three days leading up to the test.

Take a full-length practice test under timed conditions. Use the techniques and strategies you've learned in this book. Approach the test strategically, actively, and confidently.

WARNING: DO NOT take a full-length practice test if you have fewer than 48 hours left before the test. Doing so will probably exhaust you and hurt your score on the actual test. You wouldn't run a marathon the day before the real thing.

TWO DAYS BEFORE THE TEST

Go over the results of your practice test. Don't worry too much about your score, or about whether you got a specific question right or wrong. The practice test doesn't count. But do examine your performance on specific questions with an eye to how you might get through each one faster and better on the test to come.

THE NIGHT BEFORE THE TEST

DO NOT STUDY. Get together an "exam kit" containing the following items:

- A watch
- A few no. 2 pencils (pencils with slightly dull points fill the ovals better)
- Erasers
- Photo ID card
- Your admission ticket from ETS

Know exactly where you're going, exactly how you're getting there, and exactly how long it takes to get there. It's probably a good idea to visit your test center sometime before the day of the test, so that you know what to expect—what the rooms are like, how the desks are set up, and so on.

Relax the night before the test. Do the relaxation and visualization techniques. Read a good book, take a long hot shower, watch something on TV. Get a good night's sleep. Go to bed early and leave yourself extra time in the morning.

THE MORNING OF THE TEST

First, wake up. After that:

- Eat breakfast. Make it something substantial, but not anything too heavy or greasy.

- Don't drink a lot of coffee if you're not used to it. Bathroom breaks cut into your time, and too much caffeine is a bad idea.

- Dress in layers so that you can adjust to the temperature of the test room.

- Read something. Warm up your brain with a newspaper or a magazine. You shouldn't let the exam be the first thing you read that day.

- Be sure to get there early. Allow yourself extra time for traffic, mass transit delays, and/or detours.

DURING THE TEST

Don't be shaken. If you find your confidence slipping, remind yourself how well you've prepared. You know the structure of the test, you know the instructions, you've had practice with—and have learned strategies for—every question type.

If something goes really wrong, don't panic. If the test booklet is defective—two pages are stuck together or the ink has run—raise your hand and tell the proctor you need a new book. If you accidentally misgrid your answer page or put the answers in the wrong section, raise your hand and tell the proctor. He or she might be able to arrange for you to regrid your test after it's over, when it won't cost you any time.

AFTER THE TEST

You might walk out of the AP Environmental Science exam thinking that you blew it. This is a normal reaction. Lots of people—even the highest scorers—feel that way. You tend to remember the questions that stumped you, not the ones that you knew. We're positive that you will have performed well and scored your best on the exam because you followed the Kaplan strategies outlined in this section. You should rest easy if you know you did your best. If you didn't start studying until it was rather late, then you learned from your experience and will have opportunities in the future to apply improved study skills. Be confident in your preparation, and celebrate the fact that the AP Environmental Science exam is soon to be a distant memory.

| Part Two |

DIAGNOSTIC TEST

DIAGNOSTIC TEST

Time—16 Minutes
20 Questions

1. Which of the following is NOT a greenhouse gas?

   (A) Nitrous oxide
   (B) Water vapor
   (C) Sulfur dioxide
   (D) Carbon dioxide
   (E) Methane

2. Which of the following tropospheric air pollutants has been reduced significantly over the past 25 years?

   (A) Lead
   (B) Methane
   (C) Sulfur
   (D) Mercury
   (E) Ozone

3. The Clean Water Act of 1977 aims to

   (A) prohibit development on all wetlands.
   (B) eliminate discharge of pollutants into rivers and streams.
   (C) protect or restore water quality for fishing and swimming.
   (D) monitor drinking water.
   (E) do both B and C.

4. In a pyramid of biomass, there is a _____ reduction of biomass for each successive trophic level.

   (A) 10%
   (B) 30%
   (C) 50%
   (D) 70%
   (E) 90%

5. Estuaries are areas where

   (A) the beach intersects with the ocean.
   (B) waves dominate the shoreline.
   (C) fresh and salt water meet.
   (D) biodiversity is very low.
   (E) two rivers merge.

6. Intercropping refers to a growing technique in which

   (A) two or more crops are grown separately in huge plots.
   (B) two or more crops are grown at different times of the year so they do not intersect.
   (C) two or more crops are grown simultaneously in the same field.
   (D) one single type of crop is grown.
   (E) contour plowing is used.

GO ON TO THE NEXT PAGE

7. One successful technique to prevent river flooding is to

(A) channel a river with high levees adjacent to the river.

(B) allow the river to flow somewhat naturally with levees placed at a slight distance from the channel.

(C) place levees at right angles to the river flow.

(D) never use levees.

(E) create parks on floodplains.

8. The stratospheric ozone layer protects living things on Earth from

(A) ultraviolet A radiation.

(B) ultraviolet B radiation.

(C) ultraviolet C radiation.

(D) all three listed in A, B, and C.

(E) only B and C.

9. The extent of desertification is most severe in which continent?

(A) Africa

(B) Australia

(C) North America

(D) Europe

(E) Antarctica

10. The global distillation effect refers to

(A) a new moonshine technique.

(B) the deposition of pollutants in the air at the poles, far from their source.

(C) the deposition of pollutants in the air in the tropics, far from their source.

(D) the increase of air pollutants in the atmosphere.

(E) a technique of removing sulfur from coal.

11. A tsunami is caused by

(A) the convergence of two large waves.

(B) a large movement in Earth's crust, usually underwater.

(C) devastation of a coral reef near shore.

(D) all transform (horizontal) plate movement on land.

(E) movement in the molten part of Earth's core.

12. In a well-designed sanitary landfill, methane

(A) is controlled by plastic liners.

(B) escapes and adds to air pollution.

(C) is not a problem because it is used in decomposition.

(D) can be trapped and used for energy.

(E) leaks into groundwater.

13. The Ogallala Aquifer is

(A) located in the Northeastern United States.

(B) the largest groundwater deposit in the world.

(C) used for irrigation, which has lowered the water table significantly.

(D) unusable because of farm waste.

(E) both B and C.

14. The negative effects of dams include

(A) prevention of flooding upstream from the dam.

(B) loss of sediment to rebuild stream banks.

(C) facilitation of salmon migrations.

(D) increased fish populations downstream.

(E) both B and C.

GO ON TO THE NEXT PAGE

15. Salinization is caused by

    (A) salt water intrusion in the water table.
    (B) irrigation in dry climates.
    (C) traditional farming techniques.
    (D) too much rain and flooding.
    (E) oil spills.

16. The Arctic National Wildlife Refuge

    (A) is unlikely to contain more than a few days' worth of petroleum.
    (B) is a biologically robust ecosystem that recovers easily from disturbance.
    (C) has very low species richness.
    (D) is home to numerous caribou, wolves, polar bears, and arctic foxes.
    (E) is full of windmills to produce energy.

17. Scrubbers are used

    (A) to reduce sulfur emissions generated by the burning of coal.
    (B) to cleanse gas tank emissions in hybrid vehicles.
    (C) to cleanse PCB's from farmed salmon.
    (D) as part of the petroleum refinery process.
    (E) to enhance growth of wheat.

18. Synfuels

    (A) are naturally occurring liquid or gaseous fuels, including tar shales.
    (B) are alternative energies that are commonly available for consumers.
    (C) are virtually nonpolluting, so are a tremendous improvement over traditional fossil fuels.
    (D) include biogas.
    (E) are very inexpensive to produce, so will be viable for consumers in the very near future.

19. Methods for preserving genetic biodiversity include

    (A) releasing domesticated cattle into areas populated by bison.
    (B) use of inorganic fertilizers.
    (C) cultivating genetically engineered varieties of plants.
    (D) better irrigation.
    (E) farmers planting traditional varieties of crops.

20. Limiting factors

    (A) prevent growth of an organism.
    (B) may be phosphates or nitrates.
    (C) may be sunlight.
    (D) may be space.
    (E) may be all of the above.

IF YOU FINISH BEFORE TIME IS CALLED, YOU MAY CHECK YOUR WORK ON THIS SECTION ONLY. DO NOT TURN TO ANY OTHER SECTION IN THE TEST.   STOP

FREE-RESPONSE QUESTIONS

1. Fish stocks around the world are declining. Technology, overfishing, and development of coastal areas are three of the main reasons given as causes of the declining fish population. A moratorium on fishing in certain areas, such as the Georges Bank off the East Coast of the United States, has greatly changed the lives of the people who relied on fishing for their livelihoods. Many parties are worthy of blame, but fisheries biologists and fishers envision very different solutions to this problem.

   (A) (4 points) Discuss two fishing technologies that have led to the severe depletion of fish stocks.

   (B) (3 points) Describe two management strategies that need to be enacted by governments to ensure that fishing is conducted in more sustainable ways.

   (C) (2 points) Give one argument from the point of view of a fisherman and one argument from the point of view of a governmental agent about the closing of the Georges Bank.

   (D) (4 points) Name one species of seafood that we should avoid eating from the standpoint of environment sustainability. Name another species that is fished in an environmentally sound manner and is less damaging to consume. For each, explain the reason that you chose it.

2. The debate over the vast use of SUVs in the United States is clarified by looking at actual numbers of usage for a single year. Refer to the chart below to answer the following questions. (The numbers are approximate and simplified for ease of calculation.)

Data for Anytown, USA: January 1–December 31, 2006	
Number of SUVs purchased	100
Number of passenger cars purchased	160
Average number of kilometers driven by each vehicle daily (including weekends)	80
Average gas consumption per SUV	3 km per liter
Average gas consumption per sedan	8 km per liter
Number of liters per barrel	160
Number of kg of CO_2 emitted per liter of gasoline	3

   (A) (3 points) How many barrels of oil are used over the course of one year by the passenger cars that were purchased?

   (B) (5 points) How much less CO_2 would have been emitted over the course of 10 days if all SUVs had been passenger cars instead?

   (C) (2 points) Suggest two ways that we can reduce the amount of fossil fuels used.

ANSWERS AND EXPLANATIONS

1. C

Greenhouse gases, like nitrous oxide, (A), water vapor, (B), carbon dioxide, (D), and methane, (E), are those that retain heat. Sulfur dioxide (SO_2) is an industrial pollutant that is linked to acid rain, but it does not retain heat in the atmosphere and therefore is not a greenhouse gas.

2. A

Lead was used as an antiknock element in gasoline. Lead dust was released when gas was combusted. Because lead was largely phased out in the United States in 1986, the lead content in the atmosphere, as well as the lead level in the blood of all Americans, has been reduced significantly.

3. E

Drinking water, (D), is governed by the Safe Drinking Water Act. Prohibiting development of wetlands, (A), is only addressed in part by the Clean Water Act. Coastal wetlands are largely protected, but inland, seasonal wetlands are not. The elimination of the discharge of pollutants into rivers and streams, (B), and the protection or restoration of water quality for fishing and swimming, (C), summarize the two major goals of the Clean Water Act.

4. E

Biomass is an estimate of the total living material, represented by volume or weight. The pyramid shape comes from the reduction of biomass within each successive level going up the pyramid. Choice (A) is a tempting answer because each level contains only 10 percent of the total biomass of the previous level.

However, the question asks for the percent reduction of biomass, not the actual percentage of biomass for each successive level, which would be 90 percent (E).

5. C

Where the beach intersects with the ocean, (A), is called the shore. Estuaries are areas where freshwater flows downstream and meets with ocean water flowing into an enclosed bay, (C). The biota is very diverse and the water is very calm, eliminating choices (D) and (B). Choice (E), where two rivers merge, is therefore an attractive choice, but wrong.

6. C

Intercropping is a growing technique in which two or more crops are grown simultaneously in the same field, (C), eliminating choices (A), (B), and (E). This traditional technique cuts down on pest populations as insects cannot find suitable hosts directly next to each other. Monoculture is the technique in which one single type of crop is grown, (D).

7. B

Channeling a river with high levees, (A), enables building in places that shouldn't be developed because of the flood risk, but can cause flooding problems in other areas upstream. When levees are placed at a slight distance from the river channel, (B), the river can meander and a natural floodplain provides an area for excess water (eliminating choice D). Creating parks on floodplains, (E), and placing levees at right angles to the river flow, (C), don't prevent flooding.

8. E

Ultraviolet radiation is the high energy part of the electromagnetic spectrum. This radiation can be damaging to living things and is considered the primary cause of skin cancer. All UVC is absorbed by stratospheric ozone (and oxygen) as is most of the UVB. Because UVA is not absorbed, dermatologists recommend applying sunscreen every morning before going outside. Remember that tropospheric ozone is a human-produced air pollutant.

9. A

Desertification is not characteristic of Europe, (D), or Antarctica, (E). Although desertification is apparent in Australia, (B), and North America, (C), it is most severe in Africa, (A), where the Sahara and the Sahel are expanding, in part due to overgrazing in arid grasslands. This problem causes billions of dollars a year in lost agricultural productivity. In the United States, the Soil Conservation Act of 1935 was passed in response to the devastation caused by the Dust Bowl.

10. B

The global distillation effect, also called the "grasshopper effect," refers to the deposition of airborne pollutants from warmer regions in colder regions, particularly at the poles but also in other cool areas such as mountain tops.

11. B

When waves of the same wavelength come together, (A), they produce waves of greater height that have been responsible for the sinking of many ships, but these are not tsunamis. The poor health of the coral reefs, (C), in the Indian Ocean may have led to greater destruction on land after the tsunami at the end of 2004, but this is not the cause of the tsunami. Transform movement, when the plates slide past one another, on land, (D), would not cause a tsunami.

Crustal movement is closely tied to movement in the molten part of Earth's mantle, not core, (E). A tsunami is caused by a large movement in Earth's crust, usually underwater, generally due to the subsidence of one plate underneath another. Water does not compress, so once the wave is started, it continues for long distances.

12. D

In well-designed sanitary landfills, the methane can be trapped and used for energy rather than allowing it to simply escape and pollute the air, eliminating choice (B). Methane is a gas, eliminating choice (E). Although produced during decomposition by micro-organisms, it can be detrimental to the environment when allowed to into Earth's atmosphere in large quantities, so choice (C) is not correct. When rain percolates through the waste, it reacts with the methane (among other materials in the waste) and leachate is produced. Plastic liners control the leakage of leachate into the groundwater, (A).

13. E

The Ogallala Aquifer is the largest groundwater deposit in the world, (B). It is located in the Midwest and Southwest United States in the states of South Dakota, Nebraska, Kansas, Texas, Oklahoma, Wyoming, Colorado, and New Mexico, eliminating choice (A). It is not at all unusable, eliminating choice (D). In fact, it is used for irrigation, which has lowered the water table by tens of meters in some places.

14. B

The negative effects of dams include loss of sediment to rebuild stream banks. This results in two significant consequences. As this nutrient-rich sediment is no longer spread on the banks of the river, fertility of those lands is decreased. As the stream banks shrink, they no longer provide nesting areas for some birds. They prevent flooding downstream from the dam. Dams impede salmon migrations, decreasing populations. This choice is included to make sure the student is reading carefully.

15. B

Salinization is caused by irrigation in dry climates, eliminating choices (C), (D), and (E). As the water evaporates, it leaves behind salts, which cause a decrease in the fertility of the soil. Watering at the roots of a plant rather than spraying onto the surface reduces the severity of the problem. Saltwater intrusion, (A) is caused when seawater infiltrates the water table due to the removal of fresh water. Of course, irrigation may be the root cause of the saltwater intrusion.

16. D

Being a refuge means that residential or commercial uses are not allowed, even environment-friendly windmills, eliminating choice (E). It is unclear how much oil is present in the Arctic National Wildlife Refuge, but its location near the National Petroleum Reserve in Alaska suggests that the amounts could be significant, eliminating choice (A). There is very high diversity in this ecosystem that is very slow to recover from disturbance, eliminating choices (B) and (C). Environmentalists oppose drilling for several reasons: (1) The refuge is pristine; (2) it is home to numerous caribou, wolves, polar bears, arctic foxes; and (3) conservation would result in much greater energy supplies for the United States than drilling in the refuge.

17. A

Scrubbers are used to reduce sulfur emissions generated by the burning of coal. They remove virtually all of the sulfur and particulate matter in smokestacks and neutralize acid gases. They are not used in the automotive industry, (B), in the farming of fish, (C), or wheat, (E), or in petroleum refinery, (D).

18. A

Synfuels are classified as alternative fuels, as they are naturally occurring liquid or gaseous fuels or are artificially produced from plastics, rubber, or other solid waste. However, they are not commonly available for consumers, eliminating choice (B). Nor are they inexpensive, eliminating choice (E). They do not include biogas and have the same air pollution problems as other fossil fuels, eliminating choices (C) and (D).

19. E

Methods for preserving genetic biodiversity include planting traditional varieties of crops. The use of only a few genetically selected plant varieties has led to the decline of biodiversity on large corporate farms. Domesticated cattle have been problematic for wild bison, bringing diseases such as bursitis. Inorganic fertilizers and irrigation techniques encourage growing things where they might not be able to grow naturally, and so it is possible that these could increase or decrease diversity.

20. E

A limiting factor is a resource that prevents growth of an organism as a result of its scarcity. Phosphates or nitrates, sunlight, and space are all limiting factors.

ANSWERS TO FREE-RESPONSE QUESTIONS

1. This question requires a lot of specific knowledge. If you don't really know about fishing, then where could you get some points for writing something intelligent? Part C, the fishers' and wildlife biologists' arguments can be surmised from the passage. In part D, think about the fish you eat and take some guesses.

    (A) i. In the old days of fishing, the fish could "hide." They no longer can because of sophisticated sonar systems that can locate them under water.

         ii. Refrigeration techniques that permit fish to be kept for long periods of time allow for many more fish to be taken before being brought in to market. Large factory boats can stay out for months at a time.

         Other acceptable technologies: long lines on which numerous fish can be caught, trawling that disturbs the sea bed, ruining the places where fish live, and blast fishing on coral reefs.

    (B) Management strategies that have worked to improve fish stocks are being used in many places around the globe.

         i. One successful idea is to issue a limited number of permits for fishers. No additional permits are issued; they are bought and sold between interested parties.

         ii. Reduce by-catch, the wasteful collection and disposal of fish that are not the ones targeted by the expedition. Whatever the boat catches is what they have to bring to market.

         Other ideas: marine sanctuaries, closing areas where the fish stocks are depleted, preventing development in wetlands where fish spawn, not allowing fish smaller or larger than a certain size to be caught.

    (C) Argument of fishers: We need the Banks for our livelihoods; prevent development and pollution of the shorelines. Argument of fisheries biologist: Once the fish are depleted, you simply can't keep taking them or they will be fished to extinction. If they are protected for some time, the numbers will rise and then sustainable fishing can resume.

    (D) There is a wallet-sized guide for ordering seafood in restaurants. The first of three examples of seafood to avoid include farmed shrimp, which are raised in farms along mangrove swamps. The waste from the shrimp farms destroys the mangroves, a place that provides safety for many juvenile fish. The second and third common seafood choices that have been much publicized are the highly endangered Patagonian Tooth Fish (sold as Chilean Sea Bass) and shark fin soup, which requires severing the dorsal fin and throwing the shark back in the water to die.

         Wild Salmon and Halibut are both taken in a reasonably sustainable manner and most Tilapia is farmed in a sustainable way.

2. The first two parts are simple conversion problems. Think about all the numerical manipulation you made in every science or math class. Cancellations make fractions more manageable. The third part shouldn't be missed by any person, even if you haven't taken the course.

    (A) $160 \text{ cars} \times 365 \text{ days} \times \dfrac{80 \text{ km}}{\text{car}} \times \dfrac{1 \text{ liter}}{8 \text{ km}} \times \dfrac{1 \text{ barrel}}{160 \text{ liters}} = 3{,}650 \text{ barrels}$

(B) $100 \text{ SUVs} \times 10 \text{ days} \times \dfrac{80 \text{ km}}{\text{vehicle/day}} \times \dfrac{1 \text{ liter}}{8 \text{ km}} \times \dfrac{3 \text{ kg CO}_2}{1 \text{ liters}} = 80{,}000 \text{ kg CO}_2$

$100 \text{ cars} \times 10 \text{ days} \times \dfrac{80 \text{ km}}{\text{vehicle/day}} \times \dfrac{1 \text{ liter}}{3 \text{ km}} \times \dfrac{3 \text{ kg CO}_2}{1 \text{ liters}} = 30{,}000 \text{ kg CO}_2$

The difference between them is 50,000 kg CO_2.

(C)   There are so many ways we can reduce energy use. A few ideas are carpooling, taking public transportation, biking, and controlling all electricity used in the home for lights, heat, and air-conditioning.

DIAGNOSTIC TEST CORRELATION CHART

Use the results of your diagnostic test and the following table to determine which topics you need to review most. After scoring your test, check to find the area of study covered by the questions you answered incorrectly.

Area of Study	Question Number
Earth's Components and Characteristics:	
• The Structure of Earth	11
• The Role of Atmosphere	1, 8
• The Role of Water	5, 13
The Dynamic Environment	4
Dynamics and Impact of Population Growth	20
How People Use the Land	
• Agriculture	6, 15
• Forestry and Rangelands	9
• Mining and Fishing	FRQ 1
• How Populations Can Change the Land	12
• How Populations Can Save the Land	7
Energy Use	
• Energy Consumption	FRQ 2
• Fossil Fuel Resources and Use	16, 18
• Hydroelectric Power	14
Pollution and Its Effects	
• Air Pollution	2, 10, 17
• Water Pollution	3
Environmental Change and the Future	19

| Part Three |

AP
ENVIRONMENTAL
SCIENCE REVIEW

CHAPTER 3: EARTH'S COMPONENTS AND CHARACTERISTICS

IF YOU LEARN ONLY FIVE THINGS IN THIS CHAPTER . . .

1. Earth is a layered, continually moving system. At the boundaries where tectonic plates meet, earthquakes and volcanic eruptions occur, as well as the creation of mountains and deep ocean rifts.

2. Earth's climate is influenced by atmospheric circulation as well as the interaction of the atmosphere and the ocean waters.

3. Water is a vital resource on Earth that is needed for life to survive. Water moves throughout the atmosphere and crust in the hydrologic cycle.

4. Soil is a renewable resource that is formed from weathered rock material, decaying organic matter, and living organisms.

5. Soil and water are both resources that must be protected because they are susceptible to degradation and pollution. Practices such as agriculture depend on these resources but can also harm them.

From the beginning of its existence, Earth has been a dynamic, ever-changing planet. It is constantly shaping and reshaping itself. At times, the changes have been vast and vigorous, at other times imperceptible. Let's start with a brief look at the history of Earth as we explore the processes at work within it and on the surface.

THE GEOLOGIC TIME SCALE

The Geologic Time Scale divides Earth's history of 4.6 billion years into subgroups that make arranging geologic events more orderly. (See Figure 3.1.)

Eons represent the greatest span of time. There are four eons: the Hadean, Archean, Proterozoic, and Phanerozoic. The Hadean Eon represents the oldest period of time. Scientists name the time from 4.6 billion years ago to 4.0 billion years ago the Hadean. No rocks have been found from this eon that originated on Earth, although meteorites and samples of moon rocks have provided rocks of this age. The Archean Eon was from 4.0 billion years ago to 2.5 billion years ago. The oldest rocks on Earth date back to the Archean Eon. The Proterozoic ranged from 2.5 billion years ago to 542 million years ago. The Proterozoic saw an explosion of life, as the first organisms with well-developed cells appeared during this eon. We now live in the Phanerozoic Eon, which lasted from 542 million years ago to today. Organisms from this eon mainly represent the fossil record. The incredible and rapid speciation that started at the beginning of the Phanerozoic Eon, 542 million years ago, is sometimes referred to as the Cambrian Explosion. The long period of history from the formation of Earth 4.6 billion years ago until the Cambrian Explosion is sometimes informally referred to as the Precambrian.

The Phanerozoic Eon is the only eon with subdivisions. It is divided into eras (the Paleozoic Era, the Mesozoic Era, and the Cenozoic Era) which are further divided into periods. The Paleozoic Era lasted from 542 million years to 251 million years ago. The Paleozoic saw a large increase in the number of marine organisms on the planet. By the end of the era, land plants, amphibians, reptiles, and insects dominated life on Earth. The era came to an end with the largest case of mass extinction in the history of Earth. Scientists estimate that 90 percent of the species alive during this era became extinct. (See Figure 3.2.)

Era	Period	
Cenozoic	Quaternary	1.8 million years ago
	Tertiary	65 million years ago
Mesozoic	Cretaceous	145 million years ago
	Jurassic	200 million years ago
	Triassic	251 million years ago
Paleozoic	Permian	299 million years ago
	Carboniferous	360 million years ago
	Devonian	416 million years ago
	Silurian	444 million years ago
	Ordovician	488 million years ago
	Cambrian	543 million years ago

Figure 3.1 Subdivisions of the Phanerozoic Eon

Era	Period	Epoch	Millions of years ago
Cenozoic	Quaternary	Holocene	0.01
		Pleistocene	1.8
	Tertiary	Pliocene	5.3
		Miocene	23.8
		Oligocene	33.7
		Eocene	54.8
		Paleocene	65.0
Mesozoic	Cretaceous		
	Jurassic		144
	Triassic		206
	Permian		248
	Carboniferous		
	Pennsylvanian		290
	Mississippian		323
			354
Paleozoic	Devonian		
	Silurian		417
	Ordovician		443
	Cambrian		490

Figure 3.2 Geological Eras and Periods

The Mesozoic Era began 248 million years ago and is called the Age of the Reptiles. Reptiles, such as dinosaurs, dominated the land. At the end of the Mesozoic Era (corresponding with the end of the Cretaceous Period), a huge proportion of the life on Earth, around 60 to 75 percent of all species, went extinct. The immediate cause of the end-Cretaceous extinction is not certain, but evidence points to atmospheric and climatic changes caused by an asteroid strike, severe volcanic activity, or a combination of the two.

The Cenozoic Era began 65 million years ago and is also known as the Age of Mammals. With the extinction of the dinosaurs and other reptiles at the end of the Mesozoic, the mammals no longer had to compete for space or resources. This allowed the mammals to flourish.

Eras are subdivided into periods. As shown in the diagram, the Paleozoic was divided into the Cambrian, Ordovician, Silurian, Devonian, Mississippian, Pennsylvanian, and Permian. The Mesozoic is split into the Triassic, Jurassic, and Cretaceous. The Cenozoic is divided into the Tertiary and the Quaternary. The periods of the Cenozoic are further subdivided into epochs such as the Paleocene, Eocene, Oligocene, Miocene, Pliocene, Pleistocene, and Holocene. The boundaries between these geologic time intervals coincide with times where visible changes occurred on Earth. These could be times of mass extinction, evolution of species, or major geological events.

THE STRUCTURE OF EARTH

As Earth formed, dense materials sank toward the center of Earth while less dense materials rose, developing into layers (see Figure 3.3) in a process known as **planetary differentiation**.

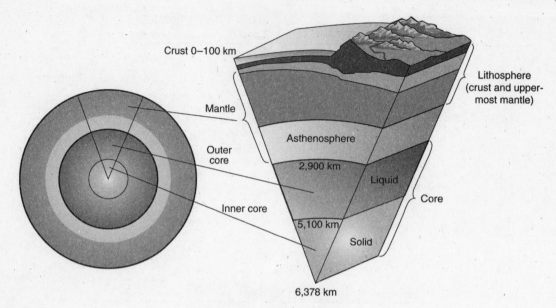

Figure 3.3 Interior of Earth
Source: Available at www.gpc.edu.

At the center of Earth is the **core**, which consists primarily of iron and is divided into two parts: the **inner core** and the **outer core**. The inner core has a radius of about 1,220 km. The temperatures (nearly 6,000°C) and pressures in this layer are so great, the iron is solid. The outer core is liquid iron, with a radius of about 3,480 km. It is believed that the movement of the liquid iron within the outer core contributes to the existence of the magnetic field on Earth.

Next is the **mantle**, a layer that makes up nearly 70 percent of Earth's interior. It is composed primarily of oxygen, silicon, and magnesium, and is also divided into two parts. Closest to the outer core is the **asthenosphere**. With temperatures that can reach nearly 4,000°C, it is the "soft," gelatinous part of the mantle. This is where **magma**, or molten rock, is formed. Above the asthenosphere, the uppermost portion of the mantle is significantly cooler (between 100°C and 1,000°C) and therefore much more solid and brittle. The outermost layer of Earth is the **crust**. The upper mantle and the crust form the **lithosphere**, which, in effect, floats on the viscous asthenosphere.

PLATE TECTONICS

The lithosphere is broken up into **tectonic plates**, or **lithospheric plates**. Each plate is roughly 100 km thick and consists of uppermost mantle with two types of crust on top: **oceanic crust** and **continental crust**. **Oceanic crust** is thin (around 5 to 10 km) but dense—rich in iron, magnesium, and silicon. The **continental crust** is thicker (around 20 to 70 km) and less dense, consisting primarily of calcium, sodium, potassium, and aluminum. Both types of crust can be found on a plate, but the denser ocean crust, as its name implies, is found below sea level. Earth has 10 major plates and many minor plates. (See Figure 3.4.)

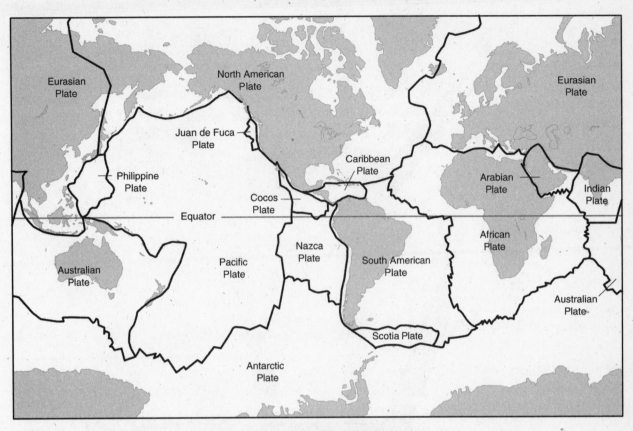

Figure 3.4 Map of Major Tectonic Plates
Source: Courtesy of the USGS

The places where plates meet are called the **plate boundaries**, and they are where earthquakes occur, volcanoes can be found, and ridges and mountain ranges form. There are three different types of boundaries: **transform, convergent**, and **divergent**.

Transform boundaries are where tectonic plates slide past each other along **transform faults**. Transform faults are a type of **strike-slip fault**, which means that the movement of the fault is horizontal. (Vertical moving faults are called **dip-slip faults**.) The movement of strike-slip faults is either sinistral (if the block on the other side of the observer moves left) or dextral (if the block on the other side of the observer moves right).

At times, movement along faults is gradual and smooth. But sometimes rocks along faults will become "stuck." If friction prevents the rocks from sliding by smoothly, pressure builds up. This pressure, or stress, builds up to the point where it suddenly snaps, creating what we know as an **earthquake**. (See Figure 3.5.)

The San Andreas Fault is a dextral strike-slip fault where the western portion of California (the Pacific Plate) is grinding past the North American Plate. This causes the earthquakes that California is known for. (See Figure 3.6.)

Buildup of pressure

Slippage (earthquake)

Figure 3.5 Earthquakes

Convergent boundaries are where two plates move toward each other. What happens at these boundaries depends on the densities of the converging plates. If an oceanic plate collides with a continental plate, the less-dense continental plate is forced up and over the oceanic plate. The oceanic plate melts as it is pushed back into the mantle under the continental plate. This is called **subduction**. At these convergent plate boundaries, also called **subduction zones** (see Figure 3.7), deep ocean trenches form where the oceanic plates subduct, and mountain ranges and volcanoes are found on the continental plate. Earthquakes are also common in these areas.

Where the oceanic Nazca Plate subducts under the continental South American Plate, the Andes Mountains have formed. This area is part of the "Pacific Ring of Fire" known for major volcanic activity and explosive earthquakes. It consists of a complex series of subduction zones between several plates, including the Nazca Plate, the Cocos Plate, the South American Plate, the Pacific Plate, the Juan de Fuca Plate, and the North American Plate.

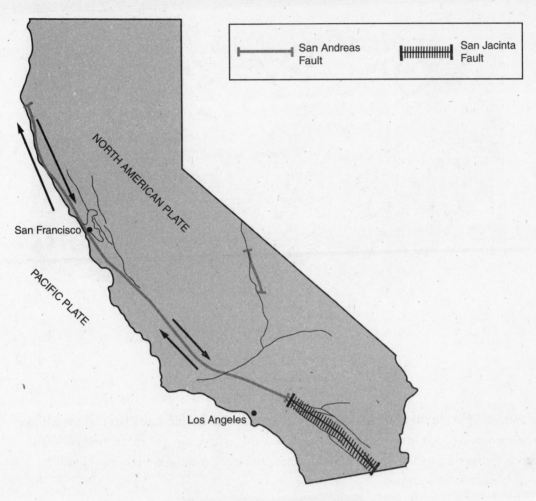

Figure 3.6 San Andreas Fault
Source: Available at http://pubs.usgs.gov.

Where two continental plates collide, one plate is forced to crumble and compress upward, or subduct into the mantle. The majestic Himalayan Mountains have formed where the Indian Plate has been forced underneath the Eurasian Plate.

When two plates with oceanic crust converge, they normally form subduction zones with deep submarine trenches. As the descending plate melts, volcanoes take shape on the ascending plate, creating an island arc, due to Earth's spherical shape. This type of boundary formed the island arc of Japan.

Divergent boundaries are where two plates move apart. This is also known as floor spreading. A rift occurs, and **magma** moves up from the mantle and cools as it reaches the surface, creating new crustal material. Although divergent boundaries do occur between continental plates, like at the East Africa Rift, this type of boundary is commonly found in Earth's oceans. This is how mid-ocean ridges develop. The Mid-Atlantic Ridge is where the North American Plate is moving away from

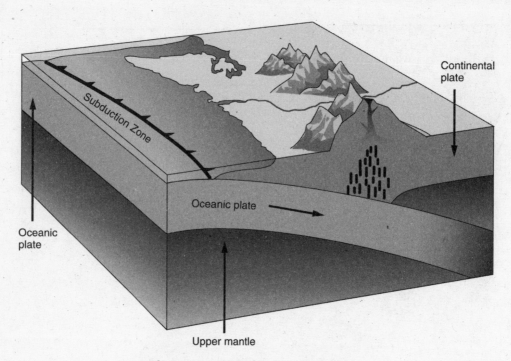

Figure 3.7 Typical Subduction Zone

Source: Available at www.usgs.gov.

the Eurasian Plate in the North Atlantic Ocean, and the South American Plate and the African Plate are moving apart in the South Atlantic. These ridges have some of the highest peaks and deepest valleys on Earth, in part because they are protected from the erosion that modifies mountains on Earth's surface.

Transform faults also occur at the ocean ridges, creating massive **fracture zones**. In these zones, the tectonic plates push the old ridges apart and the new crustal material being created rises. This creates blocky parallel structures, much like the lining of a baseball. These fracture zones are a main cause of underwater earthquakes. When earthquakes occur on the ocean floor, **tsunamis** occur. (See Figure 3.8.) They can also be created by undersea volcanic eruptions. These giant seismic sea waves (often improperly called tidal waves) can move 1,000 km/hr and can reach heights from 15 m to even 65 m.

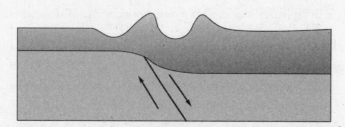

Figure 3.8 Undersea Earthquake Creates a Tsunami

EARTHQUAKES AND ENVIRONMENTAL SCIENCE

While earthquakes are a normal process on Earth, the interaction of these natural phenomena and human life can cause serious problems. Poorly built structures, overcrowding, and alteration of the environment by humans can magnify the negative impact an earthquake can have on human life and the environment.

On December 26, 2004, an earthquake originated in the Indian Ocean just off the western coast of northern Sumatra. It had the third largest magnitude ever recorded on a seismograph (between 9.1 and 9.3). The resulting tsunamis killed over 200,000 people. The coastal areas of Indonesia, Thailand, India, and Sri Lanka were most affected, though devastation from the waves occurred as far away as eastern Africa.

The earthquake and subsequent tsunamis not only took a heavy toll on human lives, but also on the environment. Many coastal ecosystems, such as coral reef, mangroves, marshes, and sand dunes, were ravaged not only by the destructive force of the tsunamis but also by the sewage and industrial waste and other pollutants distributed around the area by the waves themselves. Problems such as poisoned freshwater supplies and saltwater infiltration to arable soil are among the harmful consequences.

VOLCANOES AND ENVIRONMENTAL SCIENCE

Volcanoes are a constructive force because along with other magmatic events, they have created most of the crust on Earth. Their gaseous emissions helped create our early atmosphere and oceans over many millions of years. Volcanic eruptions create fertile soil for agriculture and create new land when lava flows into the ocean and cools.

Nevertheless, volcanoes also have destructive qualities. The obvious potential damage to human-made structures aside, it is sometimes difficult to evaluate the global impact that a volcanic eruption can have. It can release hazardous sulfur dioxide into the atmosphere. The sulfur dioxide can then combine with water vapor in the atmosphere and create sulfuric acid. Sulfuric acid interferes with solar radiation and can alter Earth's climate. One theory about the extinction of the dinosaurs at the end of the Mesozoic era was that a series of massive volcanic eruptions changed the climate and created deadly acid rain.

Also consider the fact that volcanoes often release large volumes of toxic gas and ash into the air. The eruption of Mount Tambora in Indonesia in 1815 released clouds of dust and ash high into the atmosphere. This dust and ash spread over the entire planet, blocking the Sun's rays. Sunlight, and consequently surface temperatures, decreased so much that it snowed in July in many parts of New England. The year 1815 became known as "the year without a summer." The eruption of Mount Vesuvius in 79 CE is a well-known example of the devastating effect volcanoes can have on human populations. It is now believed that the citizens of Pompeii were most likely killed by hot, toxic gas from the volcano, though the plaster casts of victims buried under mountains of ash are the most renowned images we have of the catastrophe.

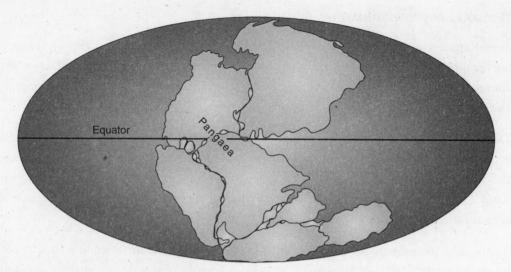

Figure 3.9 Super-Continent Pangaea

PLATE TECTONICS AND ENVIRONMENTAL SCIENCE

The continents move over long distances over millions of years. At one time, the continents were all connected in a super-continent called *Pangaea*. (See Figure 3.9.) The continents eventually broke apart and began to drift to their current locations. These changes had a great impact on the climate of the planet. It is possible that the climate changes contributed to the extinction of many organisms throughout Earth's long history.

SEASONS ON EARTH

The seasons on Earth are the result of interaction between Earth and the Sun. Remember that Earth is tilted on its axis at an angle of 23.5°, and the direction of the tilt relative to the Sun changes throughout the year. This influences the relative **altitude** of the Sun over parts of Earth. When Earth is tilted toward the Sun, the Sun is at its northernmost spot relative to the equator and daylight lasts more than 12 hours. When Earth is tilted away from the Sun, the Sun is at its southernmost location from the equator and daylight is less than 12 hours.

The Sun's place with respect to the horizon also changes. When Earth is tilted toward it, the Sun appears to be high above the horizon at noon. The noontime Sun appears lower in the sky when Earth tilts away from it.

These changes in altitude also impact the amount of energy that any point on the surface receives from the Sun. When the Sun is high in the sky overhead, the solar rays are most concentrated. As the altitude of the Sun decreases, the solar radiation becomes more spread out and less intense. (See Figure 3.10.)

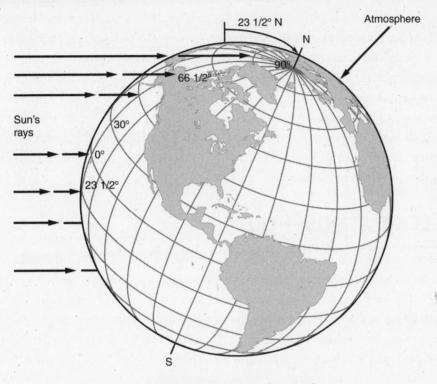

Figure 3.10 Sun and Latitude

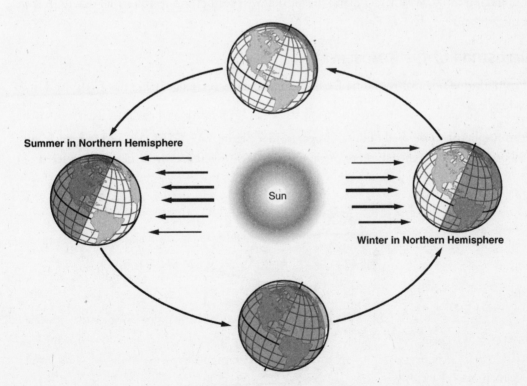

Figure 3.11 Seasons on Earth

These factors cause the changing of the seasons. For example, when the Northern Hemisphere is tilted toward the Sun, it receives extended exposure to stronger, more concentrated solar rays. The results are the spring and summer seasons, with their longer days and warmer temperatures. The reverse is happening in the Southern Hemisphere, which is tilted away from the Sun. It receives only indirect solar radiation for shorter periods of time, resulting in the fall and winter seasons. The temperatures and daylight hours at the equator, on the other hand, remain constant all year round because the equator's position relative to the Sun is not affected by Earth's tilt. In the tropics, the zone between latitudes 23°N and 23°S, the Sun is directly overhead almost year-round, so this region of Earth receives the most sunlight. (See Figure 3.11.)

THE ROLE OF ATMOSPHERE

The atmosphere on Earth has not always been the same. Scientists believe that the early atmosphere was composed primarily of hydrogen and helium. These lighter elements escaped into the upper atmosphere and out into space. There were times in Earth's history that were much more volcanically active than now. Volcanic activity added carbon, nitrogen, oxygen, sulfur, and other elements to the atmosphere. Basically all of the oxygen that we breathe was produced by photosynthesis of blue-green bacteria, algae, and green plants.

If Earth itself is a dynamic body, then the atmosphere above it is also a swirling, dynamic, ever-changing entity. Weather, climate, the air that we breathe, and winds are all due to the role that the atmosphere plays on Earth.

COMPOSITION OF THE ATMOSPHERE

The composition of air surrounding Earth is not constant. It changes from day to day and from place to place.

The atmosphere is composed of different gases. (See Figure 3.12.) Nitrogen and oxygen account for 99 percent of the composition of the atmosphere. Nitrogen is approximately 78 percent of the

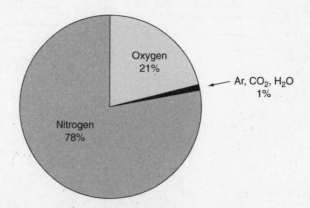

Figure 3.12 Composition of Earth's Atmosphere

atmosphere and oxygen is 21 percent. The remaining 1 percent consists of gases such as argon, carbon dioxide, water vapor, and other gases. It is important to also keep in mind that particles such as dust, volcanic ash, sea salt, dirt, and smoke are also components of our atmosphere. Some areas contain more of one of these particles than others, but these particles can have a significant influence on local weather conditions and atmospheric conditions.

The atmosphere also contains **aerosols**. Aerosols are tiny particles and droplets of liquids. Aerosols play an important role in the total amount of energy available on Earth and in the formation of precipitation in clouds.

The atmosphere consists of layers: **troposphere, stratosphere, mesosphere, thermosphere**, and **exosphere**. (See Figure 3.13.) The **troposphere** is the layer that is closest to Earth's surface. It reaches about 7 miles above Earth's surface and is the densest layer, containing nearly 90 percent of the mass of the entire atmosphere. Almost all of the clouds, carbon dioxide, water vapor, life forms,

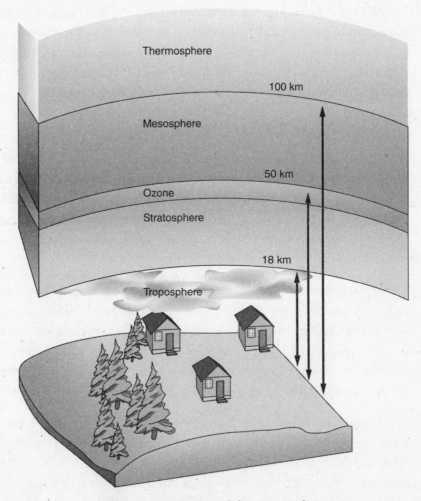

Figure 3.13 Layers of the Atmosphere

and air pollution are within the troposphere. Air moves in the troposphere in **convection cells** (see Figure 3.15) that distribute air in vertical and horizontal motions for great distances.

The air within the **stratosphere**, located about 15 miles above the surface of Earth, is very thin. The temperature grows warmer toward the top of this layer mainly because the **ozone layer** is found there. This layer of ozone absorbs ultraviolet radiation from the Sun and serves to protect life on Earth from the Sun's harmful rays.

The **mesosphere**, about 31 to 50 miles above the surface, is the coldest layer of the atmosphere. Temperature in this layer grows colder toward the top of the mesosphere, the farther away it is from the warmer stratosphere. The **mesopause** is the boundary between the mesosphere and the **thermosphere**, the fourth and last layer of Earth's atmosphere. The small amount of oxygen still present in this layer absorbs highly energetic solar radiation, which is why the thermosphere is warmer than the mesosphere. There is also a layer of ionized gases; ultraviolet radiation reacts with air particles and causes them to become electrically charged. It creates the phenomenon known as the Northern Lights. The **exosphere** is the outermost layer where the atmosphere begins to blend into space. Only the lightest gases remain, such as hydrogen and helium, with small traces of atomic oxygen.

THE GREENHOUSE EFFECT

No introduction to the atmosphere on Earth would be complete without a brief discussion of the Greenhouse Effect. Energy from the Sun enters Earth's atmosphere. But not all of that energy reaches the surface of Earth. Clouds and gases high in the atmosphere reflect about 25 percent of the energy back into space. Another 25 percent is absorbed by gases such as ozone, carbon dioxide, and methane, and water vapor.

Of the 50 percent of the energy that actually reaches the surface, some is reflected back to the atmosphere by snow, ice, and sand. The rest of the Sun's energy is absorbed by surface rocks and water, and is gradually re-released into the atmosphere. But as it is re-released, it has a longer wavelength than the original energy coming from the Sun. The gases absorb more of the longer wavelength energy than the shorter wavelength energy, and essentially hold heat in the lower atmosphere.

In itself, the Greenhouse Effect is not harmful, but a natural process that actually keeps the surface of Earth at a temperature suitable for life. Without the Greenhouse Effect, Earth's surface temperature would be −18°C to −30°C colder than it is now, about −15° to −3°C (5° to 27°F)—below the freezing point of water. Most scientists do agree, however, that too much of a Greenhouse Effect, produced by burning

AP EXPERT TIP

The Sun's light energy is absorbed by the surface, which turns the light energy into a lower form of energy: heat.

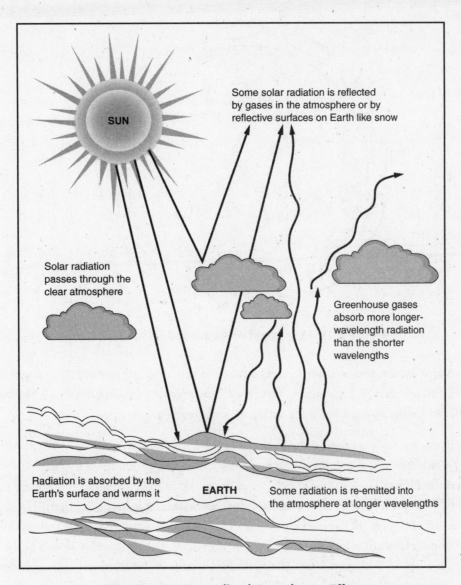

Figure 3.14 Generalized Greenhouse Effect

fossil fuels and releasing more carbon dioxide into the atmosphere, can harm the environment by trapping too much heat. (See Figure 3.14.)

ATMOSPHERIC CIRCULATION AND THE CORIOLIS EFFECT

The movement of air due to changes in air pressure is what we know as **wind**. Generally speaking, the greater the difference in pressure, the faster the wind will move. These differences in air pressure are caused by differences in the heating of the surface of Earth. Warm air from the equator rises and moves toward the poles. This creates an area of low pressure. At about 30°N and 30°S of the equator, some of the warm air begins to cool, and because cool air is heavier than warm air, it sinks,

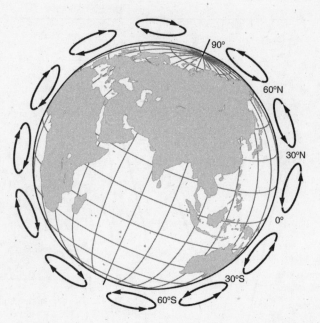

Figure 3.15 Movement of Air around Earth's Surface

creating an area of high pressure. These areas tend to be very dry. The warmer air at about 60°N and 60°S latitude also rises, creating another area of low pressure. (See Figure 3.15.) The cool air sinks back toward the equator, warming up again, and creates convection cells.

Another significant force influences the movement of the air along the surface of Earth. Wind that is moving along the surface is deflected by the rotation of Earth spinning on its axis. This is known as the **Coriolis Effect**. The Coriolis Effect causes the winds in the Northern Hemisphere to be deflected to the right relative to their origin, and it causes the winds in the Southern Hemisphere to be deflected to the left relative to their origin. As a result, the winds that are traveling north in the Northern Hemisphere will curve east, while the winds that are traveling south in the Northern Hemisphere will curve west. The opposite happens in the Southern Hemisphere: Winds that are traveling north will curve west, while winds that are traveling south will curve east. (See Figure 3.16.)

All free-flowing objects, including air and water, move to the right of their path of motion in the Northern Hemisphere, and to the left of their motion in the Southern Hemisphere. For example, surface currents in the oceans of the Northern Hemisphere turn clockwise as a result of the Coriolis Effect while surface currents in the oceans of the Southern Hemisphere turn counterclockwise. The deflection caused by the Coriolis Effect is strongest at the poles and weakest near the equator, where its influence is basically negligible. It also influences wind direction only, not wind speed.

The convection cells and the Coriolis Effect create **global wind** patterns of air circulation. (See Figure 3.17.) Winds are named for the direction from which they blow. **Polar easterlies** are created when cold air from the poles moves toward warmer low-pressure areas at 60°N and 60°S and is deflected eastward by the Coriolis Effect. Air from the high-pressure areas at the **horse latitudes** (30°N and 30°S) also move toward the low-pressure areas at 60°N and 60°S and are deflected

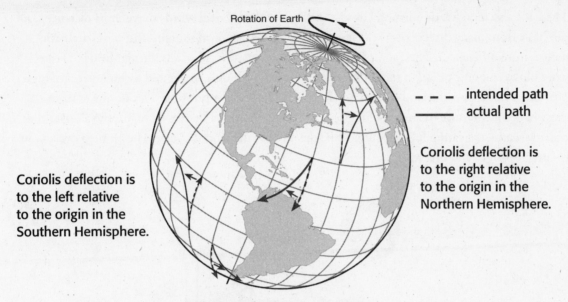

Figure 3.16 The Coriolis Effect

westward. These are called **westerlies**. The air from the high-pressure areas of 30°N and 30°S also blows toward the low-pressure areas at the equator and is also deflected westward. Explorers and traders used these winds to accelerate their trip from Europe to the Americas, and to this day they are known as the **trade winds**. The trade winds meet in an area around the equator called the **doldrums**.

Another type of wind that is found on Earth is a **jet stream**. A jet stream is a narrow band of high-speed wind that blows in the troposphere and lower stratosphere. Winds within a jet stream can reach speeds of 400 km/hr. Jet streams do not follow regular paths. Jet streams are used by airline pilots to accelerate their trips. Jet streams are also followed closely by meteorologists who trace major storms or approaching cold or warm fronts.

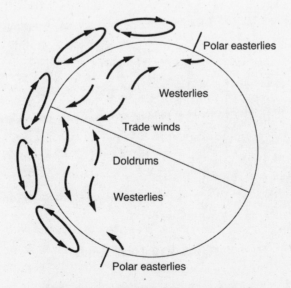

Figure 3.17 Wind Belts around Earth

There are also local winds operating on the surface of Earth. **Local winds** move short distances and can blow from many directions. (See Figure 3.18.) Shorelines or mountains can cause dramatic temperature differences. These temperature differences create local winds during the day as the wind blows from the ocean to the shore. The air over the ocean is cooler and forms an area of high pressure. The air over the land is warmer, creating an area of low pressure. The air moves from the area of low pressure to an area of high pressure, creating a sea breeze. At night, the air over the ocean is warmer, creating low pressure, while the air over the land is cooler. The air then moves out to sea, creating a land breeze.

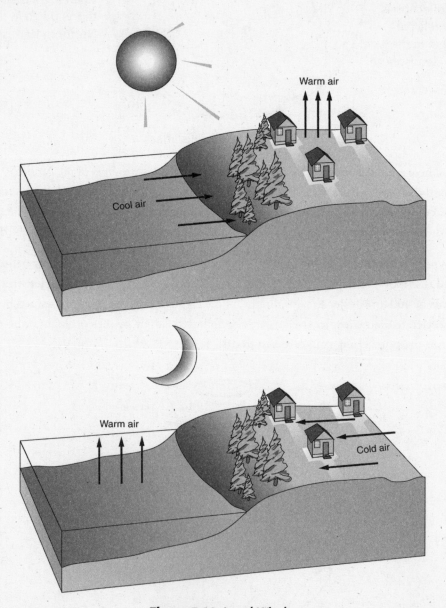

Figure 3.18 Local Winds

THE ROLE OF WATER

Water is one of the most important and vital resources on Earth. Water is everywhere on Earth: in oceans, glaciers, rivers, lakes, air, soil, and living organisms. Water is also a very disputed and conflicted resource. Approximately one-third of the population on Earth lacks clean water. A basic overview of the water on the planet will help introduce you to the concept of how precious this resource actually is.

FRESHWATER AND SALTWATER

Water on the planet constitutes the **hydrosphere** and is classified as either **freshwater** or **saltwater.** Of all the water on Earth, less than 3 percent is freshwater. Ice sheets and glaciers make up 2.15 percent of this figure and the rest—only 0.65 percent—is groundwater, in lakes and rivers, and in water vapor in the atmosphere. Freshwater, which contains less than 0.5 parts per thousand of dissolved salts, is critical to life on Earth; a vast majority of species need to consume it on a daily basis in order to remain healthy and survive. The remaining 97.2 percent of water on Earth is saltwater, which is found in the oceans. Areas where freshwater and saltwater meet are called **estuaries**.

Through the **hydrologic cycle**, saltwater from the oceans can become freshwater, and vice versa. This water cycle is powered by energy from the Sun. Water from the oceans **evaporates** into the atmosphere. Winds move this moisture around the globe until conditions are right for the formation of clouds. Eventually the moisture **condenses** into clouds and **precipitation** results. When the precipitation falls to the surface, water can **infiltrate** the ground to become part of the groundwater in the area, or it can **run off** along the surface. Some water that infiltrates the ground is used by plants that in turn release the water back into the atmosphere by a process called **transpiration.** Eventually the groundwater and surface water make their way back to the ocean, and the cycle continues. (See Figure 3.19.)

Water, particularly the ocean, also plays a role in air temperature. Water absorbs more heat than an equivalent volume of air, and takes longer to heat up and cool down, resulting in more stable temperatures near the coasts than in the interior of continents. Surface water is usually, but not always, warmer than deeper water, a result of sunlight penetration. Some areas have a **thermocline**, a layer in the water column that separates warmer surface water from cooler deep water quite abruptly.

OCEAN CIRCULATION

Water in the ocean moves in **currents**. Ocean currents are influenced by weather, the position of the continents, and the rotation of Earth. Surface currents and deep ocean currents work together to move Earth's oceans.

AP EXPERT TIP

Review pH! The measure of how acidic or basic a solution is, pH is a vital component of water chemistry. The pH value of a solution is the $-\log_{10}$ of the concentration of hydronium (H_3O^+) ions, and runs on a scale from 0 (most acidic) to 14 (most basic). Healthy freshwater usually has a pH right around neutral (7), while saltwater is slightly more basic, with a pH ranging from slightly above 7 to around 8. Aquatic life starts to suffer lasting harm at a pH around 6. Rainwater tends to be slightly acidic (pH = 5.5–6) because of absorption of CO_2 from the atmosphere.

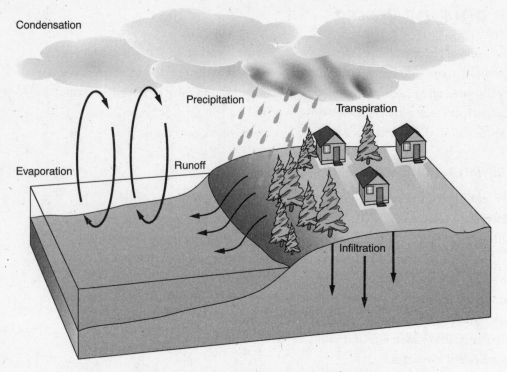

Figure 3.19 Generalized Hydrologic Cycle

Surface currents are movements of water that occur at or near the surface of the ocean. Surface currents can reach depths of several hundred meters and can reach lengths of several hundred kilometers. The Gulf Stream is one such current. Surface currents are controlled by global winds, the Coriolis Effect, and the position of the continents. These factors working together create surface currents that move in distinct patterns. Global winds, such as westerlies, polar easterlies, and trade winds, move the water in the ocean as they blow across the surface of earth. The Coriolis Effect deflects not only air but water as well. Surface currents in the Northern Hemisphere turn clockwise as a result of the Coriolis Effect, while surface currents in the Southern Hemisphere turn counterclockwise. Surface currents will deflect when they meet continents. (See Figure 3.20.)

Another factor that is important to the movement of surface currents is the differences in water temperatures. Warm-water currents originate near the equator and move warm water round the globe. Cold-water currents originate near the poles and move cold water over the globe. The result of these factors creates surface currents as shown in Figure 3.20.

Deep currents, located far below the surface of the ocean, are created by differences in density, rather than surface winds. Density of ocean water is controlled by **salinity** and temperature. Surface warm-water currents move from the equator toward the poles. The water eventually cools and forms ice near the poles, increasing the salinity of the remaining water. The increase in salinity

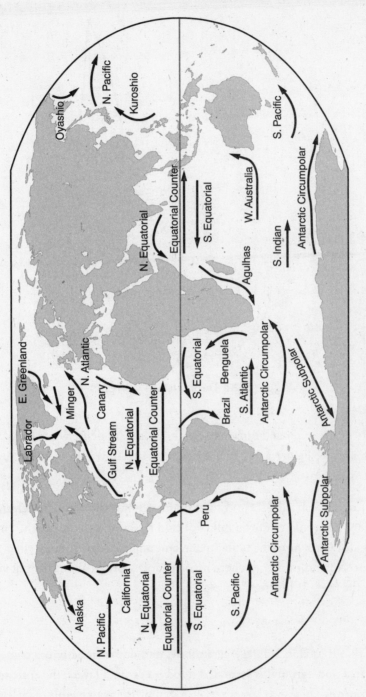

Figure 3.20 Surface Currents

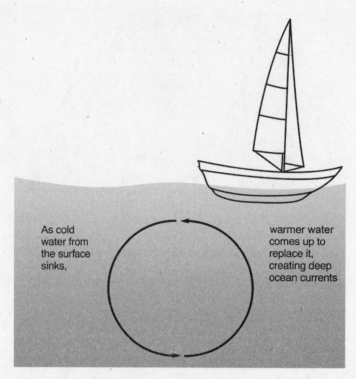

Figure 3.21 Deep Ocean Currents

increases the density of the water and causes it to sink. This cold, dense water flows along the ocean floor as a deep current. (See Figure 3.21.) These deep currents flow very slowly; a single global circuit takes hundreds of years.

Ocean circulation patterns can shift abruptly and dramatically. Let's say a large chunk of ice melts near the North Pole. This would create an influx of cold freshwater to the ocean. That water would ride along the Labrador Current to the Gulf Stream and might have a cooling effect on the weather in Europe. These sorts of abrupt changes have many people worried about the potential impact of global warming on the overall climate of the planet.

WATER USE

Much of the water that is used in industry, agriculture, and domestic activities comes from the groundwater. The area underground is separated into two distinct layers, the **zone of saturation** and the **zone of aeration**. These two layers are separated by the **water table**. The level of the water table changes with times of drought and times of heavy precipitation. (See Figure 3.22.)

Water is stored in rock layers underground called **aquifers**. An aquifer is defined by its ability to hold water and its ability to allow water to pass through it. This is known as **porosity** and **permeability**. A rock that is porous has space between the crystals or grains to hold water. A rock that is permeable has pore spaces that are interconnected, allowing water to flow easily through it. (See Figure 3.23.)

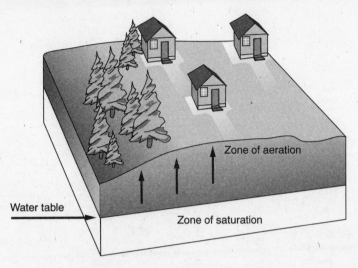

Figure 3.22 Underground Water

The porosity and permeability of a rock become particularly important when someone wants to drill a well to get water. Suppose a farmer wants to drill a well on the land in Figure 3.24. It would be best if he drilled Well A. Notice how the aquifer is confined by rock that is **impermeable** above and below. The farmer would see that his efforts would pay off. Water would flow freely from that well. However, if he tried to drill Well B, he may find that there is not much water available. That is because Well B has not been drilled deeply enough, so it sits in an impermeable layer of rock. This rock is not porous and, therefore, does not hold much water. In addition, this well is located only a short distance below the level of the water table, so it may not have had enough pressure to force water to the surface even if it did reach the permeable layer.

Water is a significant agent of weathering and erosion. This holds true for water moving underground. Groundwater is a weak acid, and certain rocks dissolve in the presence of acid. An area that has a lot of limestone, for example, is susceptible to erosion by groundwater. This is how **caves**

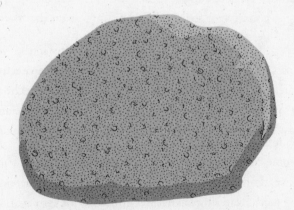

Figure 3.23 Porous Permeable Rock—Sandstone

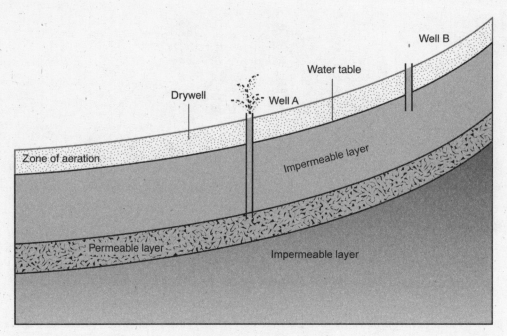

Figure 3.24 Artesian Well

are formed. They are damp because of the groundwater that flows through them, making them larger as limestone is weathered and eroded. If the water table is lower than the level of the cave, the cave is no longer supported by the water beneath it, causing the cave to collapse, creating a **sinkhole**.

WATER CONSERVATION

Clean water is a valuable resource that needs to be conserved. If we are to continue to have fresh, clean, and healthy water, we need to use water more conservatively in industrial, agricultural, and domestic situations.

Approximately 19 percent of the water used in the world is used in industry to manufacture products, cool power stations, clean products, extract minerals, and generate energy. Many industries are now trying to conserve or reduce the amount of water that they use through **recycling**.

Currently, agricultural irrigation is the single largest use of freshwater in the world. Most of the water lost to agricultural processes is lost through evaporation and runoff. New technology is being used to reduce the amount of water used in agriculture and irrigation. The new technology of **drip irrigation** brings the water directly to roots of an individual plant, using less water because the plant will directly get the water that it needs. The plant and soil will absorb the water before it can run off or evaporate.

There are many ways people can conserve water every day. Using low-flow toilets and shower heads, taking showers instead of baths, planting native plants in the yard, not letting the tap water run while brushing teeth, watering lawns after sunset to prevent evaporation, or being more economical while washing dishes can help conserve water.

THE ROLE OF SOIL

Soil is a mixture of weathered material from rock, decaying organic material, and living organisms. You could even consider the soil in your backyard to be its own ecosystem. Soil is essential to life in the biosphere and needs to be managed carefully.

THE ROCK CYCLE

The basic material that makes up soil is weathered rock. Rocks change form through the **rock cycle**. (See Figure 3.25.) Igneous rocks form from the cooling of lava (magma that has left the mantle and reached Earth's surface). The consistency of lava is known as its **viscosity**, which helps determine the flow rate of lava in a volcanic eruption and the nature of the igneous rocks that form. The igneous rocks then can be broken down into sediment through weathering and erosion. This sediment may find itself being moved through the currents of a river and then being deposited at the bottom of a quiet lake or an ocean.

If the sediment is compacted and/or cemented together, it will form a sedimentary rock. Over time, the sedimentary rock is buried and subjected to intense heat and pressure. If this is the case and it does not melt, then the rock will change into a metamorphic rock. If that metamorphic rock is subducted into the mantle and melts, it could return to Earth's surface and cool into an igneous rock.

Sedimentary rocks can change into igneous rocks or metamorphic rocks. Igneous rocks may weather into sedimentary rock, melt and cool into a different igneous rock, or transform into

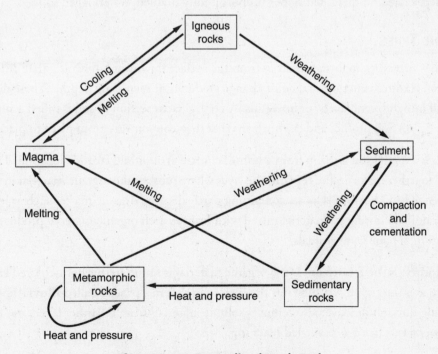

Figure 3.25 Generalized Rock Cycle

a metamorphic rock. Metamorphic rocks can change into other metamorphic rocks, weather into sedimentary rocks, or melt and become igneous rock.

SOIL FORMATION AND COMPOSITION

The formation of soil depends primarily on six factors: parent material, time, climate, plants, animals, and slope.

The source of weathered mineral matter from which soil develops is called the **parent material** of the soil. The type of parent material affects the rate of weathering, and the chemical makeup of certain rock types produces more fertile soil. If weathering has progressed over a short period of time, the parent material is a significant factor in the type of soil that forms. However, if weathering has progressed over a longer period of time, other factors are a bigger influence over the formation of the soil, such as climate. Differences in temperature and precipitation dictate whether mechanical or chemical weathering will dominate the weathering process. A hot, wet climate will produce a layer of chemically weathered soil, while a cold, dry climate will produce a thick layer of mechanically weathered debris.

Plants and animals are the primary source of organic material in the soil. Almost all soils have some organic material in them, although in varying amounts.

Slope influences soil formation and can create dramatic differences in soil type over a localized area. While steep slopes support poorly developed soils because little water can seep into the soil and erosion is accelerated on them, flat slopes support poorly drained, waterlogged soils.

MAIN SOIL TYPES

The processes that work to form soil do so from the surface downward, meaning that differences in composition, texture, structure, and color change gradually at increasing depths. These differences split the soil into different layers, or **horizons**. A vertical cross section of soil is called a **soil profile**. (See Figure 3.26.) Soil profiles and the horizons that they contain vary from place to place.

Figure 3.26 is a model soil horizon from a humid climate at a middle-latitude location. The O and A horizons together make up the **topsoil**, the layer where most of the organic material in a soil is concentrated. The O horizon is loose, containing partly decayed organic matter such as leaf litter, while the A horizon is mineral material mixed with **humus**, rich organic material produced by the decomposition of plants and animals.

The next horizon is the **E horizon**. Little organic material is found in this light-colored layer. Water percolates through it, washing out the finer materials in a process called **illuviation**. The water moving through this layer also removes soluble materials from the other layers and deposits them in lower layers in a process called **leaching**.

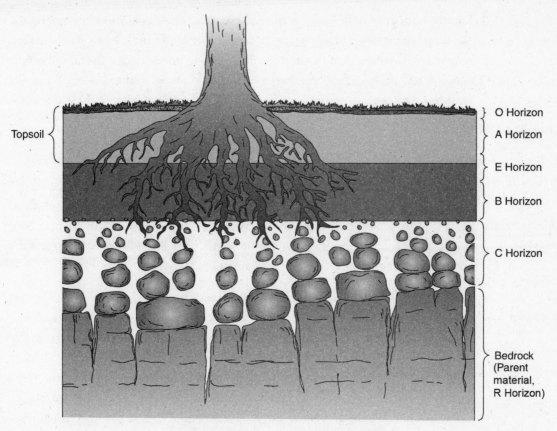

Figure 3.26 Generalized Soil Profile

The B horizon below is the area where the leached material is finally deposited. It is sometimes called the *zone of accumulation* because of that.

The C horizon sits atop the parent material or bedrock that is being weathered and contains partially altered parent material. The parent material, sometimes called the R horizon, is easily recognizable in the C horizon as large chunks or pieces.

While hundreds of soil types exist around the world, there are three very general types: pedalfers, pedocals, and laterites. **Pedalfers** contain iron oxides and aluminum-rich clays in the B horizon. Iron-rich materials and clays are leached from the E horizon into the B horizon in areas with sufficient rainfall, giving the B horizon a brown to reddish color. Pedalfer soils develop best in areas with sufficient organic material, such as forest soils. Decaying organic material provides the acidic conditions that are needed for leaching.

Pedocols are characterized by large accumulations of calcium carbonate. Pedocol soils are found in areas such as the western United States, which are drier and have grassland or brush vegetation. There is less clay in a pedocol soil than in a pedalfer soil.

Laterite soils develop in hot, wet tropical climates. These conditions are perfect for chemical weathering to occur to create deep soils. Significant leaching in these areas removes materials, such as calcite and silica-making iron and aluminum, to become concentrated in the soils. These soils are, therefore, orange and red in color. Laterites lack abundant organic matter because the bacterial action is high in these tropical regions, making for infertile soil.

PROBLEMS AND CONSERVATION

People use and abuse the soil on Earth. Approximately 40–50 percent of the land area on Earth is currently being used for agriculture, although only about one-third of this (11 percent of Earth's surface) is cropland. The rest is primarily pasture. Steep slopes, shallow soil, poor drainage, low nutrients, or high acidity or salinity prevent about four times that much land from being used for agriculture.

Land that can be used for agriculture, **arable land**, is unevenly distributed. Climate, topography, and water resources are the reasons why some lands are arable and others are not. Soil loss and soil damage are very real problems in many parts of the world. Soil is damaged by overuse from poor farming practices or overgrazing by animals. Soil that is overused loses its nutrients and can even become infertile. With no plants able to grow and to serve as protection, this soil becomes susceptible to **soil erosion**. Soil erosion is the removal of sediment or topsoil particles, usually by water or wind. Wind erosion is a particular problem in desert areas and results in a generally even loss of topsoil. In areas that are already arid, erosion coupled with climate effects can lead to serious land degradation, known as **desertification**. There are three types of erosion caused by water: splash, sheet, and gully erosion. Splash erosion refers to soil broken up by the direct impact of rain. Sheet erosion occurs when additional water, flowing in a sheet, carries away the fine particles from areas already damaged by splash erosion. Gully erosion is caused by the streams of runoff that eventually start to wear the land into gullies. Erosion feeds on itself, as degraded land is less able to support the vegetation that might prevent further erosion.

In the past, soil erosion occurred at slower rates than it does today. Human activity such as farming, logging, development, and construction accelerate the process of erosion by removing the native vegetation. Natural erosion varies from place to place and depends on conditions such as soil type, climate, slope, and type of vegetation. In many areas, soil is removed at a greater pace than new soil is formed.

Humans can do many things to avoid accelerating the rate of soil erosion, thereby protecting this renewable resource. For example, farmers can plow their fields across the slope of hills in a practice called **contour plowing**. (See Figure 3.27.) The individual rows act like a dam, preventing soil from washing away from a hillside.

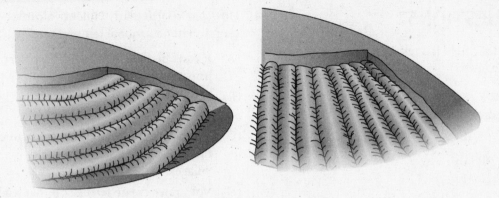

Figure 3.27 Contour Plowing

Terracing is another soil-conservation practice, where a large, steep field is converted into many smaller fields. This prevents large quantities of soil from being eroded off a large surface. The smaller fields are created to be less steep than the original field, decreasing runoff and erosion. Other farmers may employ the practice of **no-till farming** on their land. Old stalks and other plant material is left in the field after harvest in order to provide a cover from the rain, serving to reduce runoff and slow soil erosion. A specialized machine is used to "cut" a small line in the soil where seeds are planted. The cover also acts as protection for predators like snakes which will feed on pest species. **Strip cropping** is another common soil conservation method, well suited to hillsides, in which crops are planted in alternating strips arranged according to terrain and wind direction.

Nutrient depletion in soils is often as significant a problem as soil erosion. Planting only one crop in a field for years will deplete that soil of certain nutrients. Planting **cover crops**, such as clover, alfalfa, or lespedeza, between harvests helps replace the nutrients, such as nitrates, that are being depleted. Cover crops do double duty by nourishing and binding the soil together to lessen erosion.

Crop rotation, or planting different crops in a field annually, also helps reduce the nutrient depletion of the soil that can occur from planting the same crop every year.

REVIEW QUESTIONS

1. Earth's magnetic field is a result of movement within which layer?

   (A) Inner core

   (B) Outer core

   (C) Mantle

   (D) Asthenosphere

   (E) Crust

2. Which of the following is TRUE of divergent plate boundaries?

   (A) A common feature is an island arc.

   (B) They are boundaries where two plates slide past each other.

   (C) The density of the plates plays an important role.

   (D) They are where mid-ocean ridges occur.

   (E) They float on the lithosphere.

3. Most deaths in earthquakes occur because of which of the following?

   (A) Poisonous, hot gases released into the atmosphere

   (B) Violent shaking of earthquake waves

   (C) The speed with which earthquake waves travel

   (D) Giant waves called tsunamis

   (E) Poorly constructed buildings

4. How can a particularly violent volcanic eruption impact global temperature?

   (A) It can't because volcanoes have very local impacts.

   (B) The lava released from a volcano can spread over a very wide area.

   (C) The ash and dust released from an eruption can bury nearby cities and villages.

   (D) Volcanoes create fertile soil that leads to more agriculture.

   (E) Ash released from a volcanic eruption can enter the upper atmosphere where it can reflect sunlight.

5. Our current atmosphere is composed mostly of

   (A) carbon dioxide.

   (B) nitrogen.

   (C) hydrogen.

   (D) helium.

   (E) oxygen.

6. Which statement about the Greenhouse Effect is NOT TRUE?

   (A) The Greenhouse Effect is a natural process.

   (B) The Greenhouse Effect has a negative impact on the climate of the planet.

   (C) The burning of fossil fuels increases the greenhouse gases in the atmosphere.

   (D) Gases such as carbon dioxide trap solar radiation close to Earth's surface.

   (E) Most of the radiation from the Sun never even reaches Earth's surface.

7. The Coriolis Effect
    (A) affects wind direction.
    (B) affects wind speed.
    (C) is strongest at the equator.
    (D) causes winds in the Northern
        Hemisphere to move to the west.
    (E) causes winds in the Southern
        Hemisphere to move to the east.

8. Water that falls to Earth's surface may enter
   the groundwater by which process?
    (A) Transpiration
    (B) Precipitation
    (C) Infiltration
    (D) Condensation
    (E) Evaporation

9. A well that would produce water should be
   drilled into which layer?
    (A) Zone of aeration
    (B) Water table
    (C) Permeable aquifer
    (D) Impermeable aquifer
    (E) A sinkhole

10. The formation of soil is influenced by all of the
    following EXCEPT
    (A) time.
    (B) climate.
    (C) parent material.
    (D) slope.
    (E) farming.

FREE-RESPONSE QUESTION

Many places in the world are not conducive to human development because of the underlying geology. Underlying tectonics and soils are two factors to be understood when selecting an area for development. Unfortunately, many places were developed long before scientists understood the risks.

(A) (3 points) Name and describe a type of tectonically active zone where it is best to avoid dense human habitation.

(B) (3 points) Name and describe one type of soil where it is best to avoid planting crops.

(C) (2 points) Name two specific places in the world where geological disasters have occurred in the past 2,000 years.

(D) (2 points) If you were on a world development council, keeping in mind the limitations of financial resources, recommend two ideas that will prevent the repeat of past geological problems.

ANSWERS AND EXPLANATIONS

1. B

The inner core, (A), and the crust, (E), are both solid and, therefore, will have no movement, eliminating them right away. Parts of the mantle, (C), are fluid in nature, but it is not made of the iron needed to influence or create a magnetic field. The asthenosphere, (D), is the mushy layer within the mantle where convection occurs.

2. D

Island arcs, (A), are common features of convergent boundaries where two ocean plates meet. The boundary where two plates slide past each other, (B), is a transform boundary. The density of the plates, (C), is important at subduction zones, where the denser plate sinks below the less-dense plate, which is also a feature of convergent plates. Tectonic plates are part of the lithosphere, which floats on the asthenosphere.

3. E

Volcanoes, not earthquakes, can release hot gases, (A). While it is true that some underwater earthquakes create giant killer waves called tsunamis, not all earthquakes do. Therefore, choice (D) does not apply to earthquakes in general. Choices (B) and (C) are not correct because, although these are factors in an earthquake, if people were not tightly packed together in poorly constructed homes upon potentially unstable soils, the impact of an earthquake on people would be minimal.

4. E

Volcanoes formed the early atmosphere of the planet, so they definitely can have global impact. Choices (B) and (C) can be eliminated because their effects are minimal and localized in the area nearest the volcano. While (D) is true, an increase in agriculture

does not signify a significant change in global temperature.

5. B

Hydrogen, (C), and helium, (D), were the main components of the early atmosphere on Earth, but most of these gases floated off into space. Choice (A) is incorrect because life can't survive in an atmosphere of carbon dioxide. Oxygen, (E), may seem like a logical answer because life needs oxygen. But oxygen is only 21 percent of the atmosphere, while nitrogen is 78 percent.

6. B

The Greenhouse Effect does occur naturally and is responsible for keeping the temperature on the planet at a level that can support life, (A). It is true that human activity, including the burning of fossil fuels, can increase greenhouse gases, such as carbon dioxide, that trap heat close to the surface. Nearly 50 percent of the incoming solar radiation is reflected back into space or absorbed by the clouds in the upper atmosphere. This means that choice (E) is true.

7. A

Choice (B) can be eliminated because the Coriolis Effect does not impact the speed of wind, only the direction. Choices (D) and (E) are incorrect because the movements have been reversed. Winds in the Northern Hemisphere move to the east while winds in the Southern Hemisphere move to the west. The effect of the Coriolis Effect is greatest at the poles. This eliminates answer (C).

8. C

Transpiration is the release of water into the atmosphere by plants and photosynthesis. Water falling to the ground in the form of rain, snow, or ice is precipitation, so choice (B) is wrong.

Condensation forms clouds and evaporation causes water vapor to enter the atmosphere. This eliminates (D) and (E).

9. C

Answer (A) is incorrect because there is no water in the pore space of the zone of aeration. The water table is the boundary between the zone of aeration and the zone of saturation. An impermeable layer of rock does not have the pore space to hold much water. A sinkhole, answer (E), is not a place to drill a well.

10. E

Soil forms as a rock (parent material) and weathers over time. This means that (A) and (C) do impact soil formation. Because a warm wet climate can speed up weathering, (B) is not the answer you are looking for. If soil is being formed on a steep slope, the soil will be very thin. So slope and gravity both play a role in soil formation.

ANSWER TO FREE-RESPONSE QUESTION

(A) Any place that is close to a plate boundary can be tectonically active, subject to earthquakes and volcanic eruptions. Convergent zones, where plates move toward each other; divergent zones, where plates drift apart; or transform boundaries, where plates move side by side, are all acceptable answers.

(B) Oxisols, which are tropical and subtropical soils, are very poor in nutrients because leaf litter decomposes quickly. When trees in the rain forest are cut and crops are planted or animals are grazed, the soil quickly becomes highly infertile. Another good choice are aridosols, or desert soils, that can support some grazing but currently are being overused. Almost any other type of soil could be used in this answer if it is accompanied by a good explanation.

(C) The tsunami of 2004 in the Indian Ocean is a recent catastrophe. Events that most students should have heard of are earthquakes in California, Iran, Turkey, Japan, and the volcanic eruptions of Mt. St. Helens in Washington and Mt. Pinatubo in the Phillipines.

(D) The first major idea is the enforcement of building codes. Many deaths were caused in the Turkish earthquake because of disregard for the measures that must be taken by building developers to minimize deaths. Deaths have been prevented in several significant earthquakes in highly populated areas such as in California and Japan. It is also most advisable for people to avoid settling in low-lying areas, but the question reminds us that poverty may make this impossible. Therefore, the tsunami warning system is an essential tool needed to save lives. Hotels at beach resorts should be built on higher ground.

CHAPTER 4: THE DYNAMIC ENVIRONMENT

IF YOU LEARN ONLY FIVE THINGS IN THIS CHAPTER . . .

1. Ecosystems consist of nonliving and living factors that interact to allow the ecosystem to respond to environmental changes. Nonliving factors include physical and chemical conditions (e.g., temperature, rainfall, soil composition, etc.). The living factors include organisms, species, populations, and communities. The living factors interact in complex ways.

2. Ecosystems are driven by energy that comes from the Sun and other sources. Energy and matter flow together through the living portions of the ecosystem. Energy flows one way only, but matter gets recycled within and between ecosystems.

3. Ecosystems can vary in their species diversity, the number and abundance of different species. Natural selection and evolution and movement of species can bring about species diversity. Diverse ecosystems are believed to be more resistant to environmental changes.

4. Ecosystems are not static. They change constantly and must be able to adapt to environmental stresses (e.g., climate shifts, pollution, habitat destruction). Several mechanisms allow ecosystems to adapt, including movement of species (organisms, species, populations), natural selection/evolution (species, populations), and ecological succession (communities, ecosystems).

5. Biologically important elements (carbon, oxygen, nitrogen, water, sulfur, phosphorus) are cycled through living and nonliving components of ecosystems, both locally and globally. These cycles have important implications for living things and for the planet. These cycles can be and are being altered significantly by human activities.

ECOSYSTEM STRUCTURE

The living world is composed of many ecosystems that interact and affect each other on a global scale. For example, the tropical rain forests take up carbon dioxide, a heat-absorbing or greenhouse gas, and by doing so help influence the global temperature. To define ecosystem, let's review the parts and consider an example of one.

Consider a salt marsh or tidal estuary found along the coastal United States. This wetland lies where a river or stream empties directly into a bay, sound, or the ocean. Here, freshwater meets saltwater, and the water changes twice daily with the ebb and flow of the tides. The salt marsh is generally shallow and muddy. Mud flats can be exposed to the air during low tides and flooded during high tides. The temperature of the water can be warm or cool depending upon the time of day, the tide, and the season of the year. The salinity or salt concentration of the water varies with the tide, but it is usually lower near the mouth of the river or stream and becomes higher as you move farther out near the open ocean. The mix of salt water and freshwater is known as **brackish water**.

All ecosystems have two main components:

1. **Abiotic factors**—"nonliving" factors such as physical or chemical conditions in the environment. In the salt marsh example, abiotic factors would include water temperature, salinity, pH, soil composition, oxygen content of the water, and mud.

2. **Biotic factors**—"living" factors such as all of the living organisms within that environment. In our salt marsh, biotic factors would include all of the plants (marsh grass, shrubs), animals (fish, worms, shellfish, crabs, insects, birds), and microorganisms (bacteria, plankton) that live in the water, in and on the mud flats, and on the shoreline.

The biotic and abiotic factors in an ecosystem interact to allow the system to respond to changing environmental conditions.

We can further characterize biotic factors of an ecosystem into a hierarchy from lowest level to highest level:

1. **Organisms**—These are individual life forms. For example, individual fiddler crabs, flounder, or marsh grass plants found in the salt marsh example.

2. **Species**—Groups of organisms that share similar appearances, behaviors, chemistries, and genetic structures. In the salt marsh, the fiddler crabs, fish, and birds are separate species.

3. **Population**—Groups of the same species living in the same place at the same time. The fiddler crabs in the salt marsh example are one population, which is different from fiddler crabs of the same species that live in a salt marsh two miles north of our example one.

4. **Community**—Populations of different species that live in the same place at the same time and interact with each other. In the salt marsh example, the fiddler crabs, fish, birds, and plants represent a community.

So, finally, we can define an **ecosystem** as communities of different species that interact with each other and with the abiotic factors, allowing the overall system to respond to environmental changes.

An important part of the definition of an ecosystem is that the species within the community interact with each other. Each species has a role within the community. Sometimes those roles overlap with those of another species, and those two species then interact. Let's examine the roles of species and then look at how different species can interact with each other in a community.

NICHE: EACH SPECIES IN AN ECOSYSTEM HAS A DIFFERENT ROLE TO PLAY

Every species within an ecosystem has a functional role in that system called a **niche**. A *niche* consists of all the physical, chemical, and biological conditions that a particular species needs to live and reproduce within the ecosystem. In a salt marsh, various species of mussels may be found in different parts of the marsh. They feed by filtering, thereby removing substances, including toxins, from the water. They are also food for various birds. However, one species of mussel may occupy a particular niche near the mouth of the stream because the salinity is low, while a different mussel species might be found in a different niche closer to the bay side where the salinity is higher. While both species perform similar roles in the ecosystem, the physical factor of salinity separates one niche from another. Other factors that can distinguish one niche from another include temperature, oxygen content, type of food resources available, and so on.

Generally, species can be classified into two groups based on the niches that they occupy:

1. **Specialist species**—a species that can occupy only a few niches. For example, they may have a narrow tolerance for salinity or temperature changes. Perhaps, they rely on only one type of food source.

2. **Generalist species**—a species that can occupy a broad range of niches. For example, a crab that can burrow in the mud where oxygen content is low and yet move about in the open water channels where oxygen content is high. Alternatively, the crab may be able to use a variety of food sources, such as hunting small fish or eating decayed material from the mud surfaces.

There are advantages and disadvantages to being either a generalist or a specialist depending upon the status of the environment. In a stable environment, a specialist may have an advantage because its narrow niches have few competitors, but in a rapidly changing environment, a generalist can usually adapt better than a specialist and survive. Specialists are prone to extinction, while common or invasive species tend to be generalists.

AP EXPERT TIP

The Competitive Exclusion Principle says that two species will never occupy the same niche. If they do, competition will occur and one species will be excluded.

AP EXPERT TIP

Examples of specialist species are the panda, which eats only bamboo, or the Kirkland's warbler, which nests only in young Jack Pine forests.

INTERACTIONS BETWEEN SPECIES

Species within communities can interact in the following three ways:

1. Competition for limited resources
2. Predation
3. Symbiosis
   - Parasitism
   - Mutualism
   - Commensalism

COMPETITION

Within an ecosystem, many species often compete for limited resources such as food, sunlight, space, and water. One species may win out over another by one of three ways:

1. **Interference**—One species limits the access of another species to the resource. In the salt marsh, a fiddler crab might defend its mud hole from other organisms, or a seagull might defend its scavenged crab from all other birds.

2. **Exploitation**—Two or more species have equal access to the resource, but one uses it more quickly or efficiently than the others, thereby gaining an advantage. For example, as the tide strands small fish in pools on the mud flats, flying birds can cover the area faster than blue crabs, thereby taking greater advantage of the resource.

3. **Resource partitioning**—Two or more species have equal access to the resource, but use it at different times or places or in different ways. For example, two species of birds in the marsh might hunt for worms and mollusks in the mud, but one bird has a longer bill and can feed on worms that live deeper in the mud than can the other bird. Therefore, both bird species can coexist when using the same resource.

No two species can occupy the same niche at the same time in the same area. This is called the **exclusion principle**. One of the competing species will move to another area, change its behaviors, lose members of its population, or become extinct.

PREDATION

One of the most obvious ways that species can interact is that one species, the predator, eats another species, the prey. There are two obvious types of predator-prey

> **AP EXPERT TIP**
>
> No species benefits from competition. It is considered a negative situation for both because they have to expend valuable energy and risk being hurt.

relationships. In a **carnivore-prey** relationship, the carnivore hunts and eats its prey. For example, an egret hunts and eats fish as it moves through the marsh. In an **herbivore-plant** relationship, the herbivore (e.g., locust, cow, horse) grazes on the plants in an ecosystem.

To elude predators, prey have evolved several strategies, such as protective shells or casings, exuding toxins, running or flying faster than the predator, camouflage, hiding, being active at different times than predators, and mimicking other objects (rocks, vegetation, predators).

SYMBIOSIS

Symbiosis is a different form of species interaction in which one species, the **symbiote**, lives on or inside another species, the **host**, for most of its life cycle. The type of symbiotic relationship depends upon who benefits and who gets hurt in the relationship:

- **Parasitism**—In this relationship, the symbiote benefits, while the host gets injured or perhaps even killed. One might even consider this a specialized form of the predator-prey relationship. Some examples of parasitism include tapeworms, lice, mistletoe plants, and leeches.

- **Mutualism**—In this relationship, both the symbiote and the host benefit from the relationship. For example, insects that feed on nectar from flowers not only receive food, but they also exchange pollen among the flowers, thereby helping the plants reproduce. Bacteria in the roots of legume plants help fix nitrogen from the soil for use by the plants and benefit from the plants' nutrients.

- **Commensalism**—Here, one species benefits while the other species neither benefits nor gets harmed. Some land plants called epiphytes live on trees where they get water or nutrients from tree bark surfaces. The host trees are not harmed, nor do they benefit from the presence of the epiphytes.

Interactions between species within ecosystems are essential to the functioning of the ecosystem and to the services that the ecosystem provides.

ARE ALL SPECIES IN AN ECOSYSTEM EQUAL?

The species in any ecosystem can be categorized in the following ways:

- **Native species**—These species normally live within a given ecosystem. In the salt marsh example, marsh grass, fiddler crabs, killifish, snails, and the various birds are all native species.

> **AP EXPERT TIP**
>
> Species benefit from predation—even the prey! The predators eliminate the weak, old, or least adapted individuals because they are easier to catch. Those sick or least adapted individuals will not be able to reproduce and pass on their "weak" genes to the next generation, and competition between individuals is reduced.

- **Alien, invader, or introduced species**—These species are not normally found within a given ecosystem, but they are either deliberately or accidentally introduced into the ecosystem, usually by human activities. Once in the ecosystem, they may thrive and either benefit or hurt the ecosystem when they outcompete native species. For example, the zebra mussel was accidentally introduced into the Great Lakes region of the United States when foreign freighters dumped bilge water containing zebra mussel larvae. The zebra mussels had no natural predators in the new ecosystem and thrived. They now present huge problems by fouling intakes to power plants, ports, and other industries on the Great Lakes.

- **Indicator species**—These species are native species whose abundance and success provide information as to the overall health of the ecosystem. Their decline serves as an early warning that the ecosystem is in danger. Frogs are a good indicator species in aquatic ecosystems because their permeable skin makes them more sensitive to pollutants and toxins than other organisms.

- **Keystone species**—These species play one or more important roles that affect many other species within an ecosystem. In effect, the keystone species control the structure of the community within an ecosystem. The loss of the keystone species could lead to the decline and extinction of many other species within the ecosystem. For example, sea otters off the coast of California feed on sea urchins that, in turn, feed on the large kelp forests that provide food for other animals (mollusks, crustaceans, fish). If the sea otter gets taken out of the ecosystem, the sea urchins would multiply and destroy the kelp beds, thereby causing declines in the populations of these other animals. Thus, the sea otter plays a keystone role in that ecosystem. The role of an organism as a keystone may not be obvious and may remain unknown until it has been removed from an ecosystem.

Some ecologists argue that all species within an ecosystem have equally important roles, but many scientists favor the keystone species concept.

DIVERSITY AND EDGE EFFECTS

The structure of the biological community within an ecosystem is important in describing the age of the ecosystem as well as its ability to withstand changes. The structure of the community is best described by its **species diversity**, which is the number of different species and their relative abundances in a given area. Species diversity increases as the number of species increases and/or as the number of individuals in the total population becomes more equally distributed among the species. As communities age, they tend to get more diverse. This diversity may help the ecosystem adapt to changes. Ecosystems with more diversity may be able to withstand environmental challenges better than those with little diversity. While this notion seems logical, it is unproven. For example, redwood forests have great species diversity and can be resistant to changes. But if a significant amount of the redwood forest is destroyed, the ecosystem may never recover. In contrast, grasslands have little diversity but can recover quickly when significantly damaged. So, it remains unclear exactly how species diversity contributes to ecosystem stability.

Biological communities can interact when they come into contact at boundaries called **edges**. Edges can result from natural boundaries (abrupt changes in soil or rock types), natural disturbances (fires, floods), or human activities (agriculture, land clearing, livestock grazing, timber clearing, roads). Usually, there are transitional zones at edges called **ecotones**. In the ecotone, each community competes with the other for resources and a new community called an **edge community** forms.

As illustrated in Figure 4.1, the following types of edges are possible:

- **Abrupt edge**—abrupt changes between communities with no edge community (Figure 4.1A). Abrupt edges can be caused by changes in rock and soil types, topography, or microclimate.

- **Mixed edges**—species from both communities invade the ecotone and compete for resources (Figure 4.1B–D).

  — Dominant mixed edges—species from one community or the other dominate the edge community (Figure 4.1B–C).

  — Nondominant mixed—species from both communities inhabit the ecotone equally (Figure 4.1D).

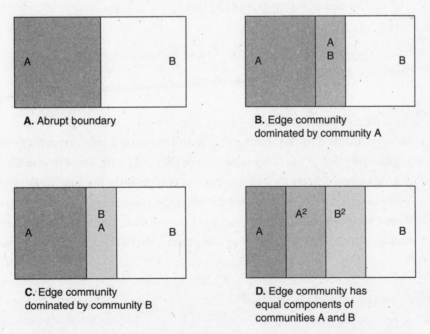

A. Abrupt boundary

B. Edge community dominated by community A

C. Edge community dominated by community B

D. Edge community has equal components of communities A and B

Figure 4.1 Types of Ecotones Between Two Communities, A and B

AP EXPERT TIP

White tail deer are an example of an "edge species" because they like to eat the new growth of buds and berries on the edge of a grassy ecosystem and a forest.

Edge communities usually have more diversity than the communities on either side of the ecotone. This phenomenon is called the **edge effect**.

Edges of communities and ecotones have different and, often, more extreme environmental conditions than those in the surrounding communities. For instance, in the edge between a field and a forest, trees in the edge community have more exposure to the Sun, wind, and rain than those in the forest community. Similarly, shrubs and grasses in the edge community have less exposure to these conditions than those in the field community.

MAJOR BIOMES

European ecologists recognized definite patterns in plant communities across the globe and classified these patterns into **biomes**. A biome is a distinctive combination of plants and animals characterized by a uniform type of vegetation. There are nine major biomes and each varies with respect to temperature, rainfall, growing season, soil type, etc. The major biomes are listed in Table 4.1.

If you look at a map of the world biomes, you will see the following generally ordered five-step pattern as you go from the equator to the poles:

1. Tropical forests (equatorial regions)
2. Deciduous forests (above the tropics)
3. Taiga
4. Tundra
5. Ice and snow (polar regions)

This pattern exists because it generally gets colder and wetter as you move from the equator into the temperate zones, and then, once when you reach the polar regions, precipitation falls mainly in the form of snow. You can also see the same pattern when you climb a mountain. As you go from the base to the summit, you will pass from deciduous forests through taiga and tundra until you reach the snowcapped peak. This pattern exists because the climate changes that occur with changes in altitude are similar to the pattern that occurs with changes in latitude.

Table 4.1 Nine Major Terrestrial Biomes of the World

Biome	Climate	Vegetation	Animals
1. Tundra	Low precipitation, mostly snow Subfreezing or freezing temperatures Permafrost soil, short growing season	Low-growing plants (moss, lichens)	Mammals (hares, voles, wolves, caribou), birds (owls, migratory birds)
2. Taiga (boreal or coniferous forests)	Long, cold, dry winters and short, moderately warm, wet summers	Evergreen trees (spruce, firs), few deciduous trees (aspen, birch)	Mammals (moose, beaver, bears, wolves), birds (owls, migratory birds)
3. Temperate Deciduous Forest	Moderate temperatures with four seasons. Warm summers, cold, snowy winters, abundant precipitation	Deciduous trees (oak, hickory, maple, poplar, sycamore, beech)	Mammals (foxes, deer, wolves, hares), birds (owls, hawks, migratory birds)
4. Temperate Rain Forest	Moderate temperatures, abundant rainfall and moisture	Conifers (spruce, fir, redwoods), broadleaf evergreens	Similar to temperate deciduous forests
5. Tropical Rain Forest	Constant warm temperatures with heavy rainfall and long growing seasons	Large, broadleaf evergreen trees, diverse plant life, many species	Many species of mammals, birds, insects, reptiles
6. Savanna	Tropical wet and dry seasons	Grasses, shrubs, and small trees	Grazing mammals (zebras, wildebeests, giraffes, antelopes, elephants), predatory mammals (lions, cheetahs, hyenas), birds (migratory)
7. Temperate Grasslands	Hot, dry summers and cold snowy winters	Tall and short grasses	Mammals (cattle, sheep, buffalo)
8. Shrubland	Hot, dry summers and moist winters	Shrubs	Mammals, reptiles, birds
9. Desert	Little precipitation (hot or cold)	Shrubs, cacti	Reptiles, mammals adapted to dry conditions (rodents)

FLOW OF ENERGY IN ECOSYSTEMS

All organisms require energy to carry out various biological functions. In an ecosystem, energy from a source gets captured by one group of organisms and gets transferred to other organisms either when they are eaten or when they die. The organisms that capture the energy from the source are called **producers** or **autotrophs**. Producers utilize this energy to make food from materials in the environment.

For most ecosystems, the primary energy source is the Sun. Most producers in terrestrial and aquatic ecosystems (plants, algae, phytoplankton, cyanobacteria) capture sunlight in a process called **photosynthesis**. In photosynthesis, these organisms use light energy to make food (sugars, starches) from carbon dioxide and water and make a by-product, oxygen. In some ecosystems, such as hydrothermal vents deep in the ocean, the energy comes from inside Earth (i.e., geothermal energy). In such cases, **chemosynthesis** is used to create organic matter. The source of energy is not sunlight, but a 1-carbon molecule like carbon dioxide or methane.

The food that producers make not only provides energy for them, but also provides energy for other organisms that eat them (**consumers** or **heterotrophs**). In addition to food, producers provide the oxygen that makes up a substantial portion of Earth's atmosphere. Let's take a closer look at the process of photosynthesis.

PHOTOSYNTHESIS

Photosynthesis occurs in the chloroplasts of cells in the leaves of plants and certain single-celled eukaryotes, or in the membranes of photosynthetic bacteria. Here is a simple chemical equation to give a better picture of what is going on:

$$6\ CO_2 + 6\ H_2O \xrightarrow[\text{chlorophyll}]{\text{Light}} C_6H_{12}O_6 + 6\ O_2$$

carbon dioxide   water   glucose*   oxygen

*Although glucose is commonly used in this equation, the products are mostly smaller (i.e., 3-carbon) sugar-phosphate molecules that are used to make more complex sugars and starches.

Photosynthesis actually involves many chemical reactions, not just one. (See Figure 4.2.) These reactions can be divided into the following two major categories: the light reactions (occurring in the thylakoid membrane of the chloroplast or the

---

AP EXPERT TIP

The inorganic molecules of CO_2 and H_2O are low energy molecules (we know molecules are low in energy if they do not burn). During photosynthesis, these low energy molecules are used to make a high energy, organic molecule like $C_6H_{12}O_6$.

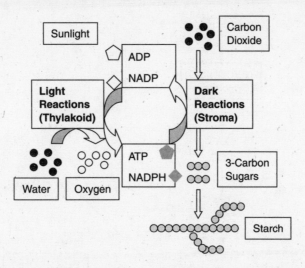

Figure 4.2 Summary Diagram of the Process of Photosynthesis

cell membrane of cyanobacteria) and the dark reactions (occurring in the stroma of the chloroplast or the cytoplasm of cyanobacteria).

1. In the light reactions, the following occurs:

   - Energy from sunlight is captured by chlorophyll and stored in **adenosine triphosphate** (ATP).

   - Electrons are passed from chlorophyll to an electron carrier, nicotinamide-dinucleotide phosphate (NADPH, electron-loaded form).

   - The electrons that were lost from chlorophyll are replaced with ones from water. Water gets split to make oxygen, which is ultimately released into the atmosphere.

2. In the dark reactions, the following occurs:

   - In a cycle of chemical reactions, carbon dioxide gets fixed from a gas into a solid molecule to make 3-carbon sugars.

   - These simple 3-carbon sugars are used to make complex sugars and starches.

   - The energy and electrons needed to make these sugars come from the ATP and NADPH made in the light reactions.

The sugars and starches are stored in the cells and tissues of the photosynthetic organisms.

The energy stored in the food made by producers gets transferred to consumers (animals) when they eat the producers; some consumers eat other consumers as well. *Producers* are organisms that utilize energy to make food. They include photosynthetic organisms, such as some bacteria, algae, and plants. When both producers and consumers die, organisms called **decomposers** (e.g.,

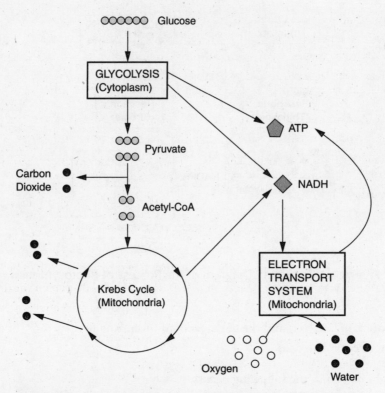

Figure 4.3 Summary Diagram of the Process of Respiration

bacteria, fungi) break down the tissues of the dead organisms to extract energy. But how do consumers extract energy from food? They do so by a process called **respiration**. Let's take a closer look at this process.

RESPIRATION

Like photosynthesis, respiration can be represented by a simple chemical equation that is the opposite of photosynthesis:

$$C_6H_{12}O_6 + 6\ O_2 \text{------------------>}\ 6\ CO_2 + 6\ H_2O$$

glucose   oxygen   carbon dioxide   water

Cells "burn" sugar in a controlled way and harvest the energy to make ATP. The process of respiration is a complex one that occurs in the following three stages (see Figure 4.3):

1. **Glycolysis**—In this 10-step pathway that occurs in the cytoplasm, a molecule of glucose (6-carbon) is ultimately broken down into two pyruvate molecules (3-carbon), four ATP, and some electrons bound to **nicotinamide-dinucleotide** (NADH, an electron carrier).

2. **Krebs Cycle** or **Citric Acid Cycle**—This process occurs in the matrix of the mitochondria or cytoplasm of a prokaryotic cell. Upon entering the mitochondrion, pyruvate gets stripped

of a carbon atom to make acetyl-CoA (2-carbon), some electrons are removed to NADH, and a molecule of carbon dioxide is released. Acetyl-CoA gets completely broken down into carbon dioxide in a series of steps, more electrons get transferred to electron carriers, and two more ATP get made. Furthermore, this process only happens if oxygen is present (i.e., aerobic respiration).

3. **Oxidative Phosphorylation** or **Electron Transport Phosphorylation**—All of the electron carriers from glycolysis and the Krebs cycle transfer their electrons to oxygen through a chain of proteins in the inner membrane of the mitochondria (or the cell membrane of prokaryotic cells). As the electrons get transported, protons move across the membrane to produce ATP and to combine with oxygen to yield water. Overall, the cell gets 38 ATP for every glucose molecule that goes through respiration.

Why do cells go through such a complex process to break down food? By doing these steps, they retrieve energy in a usable form (ATP) rather than 100 percent heat energy. However, the process is not 100 percent efficient, and some of the energy from glucose escapes as heat.

FOOD CHAINS, FOOD WEBS, AND TROPHIC LEVELS

Producers, consumers, and decomposers are organized into food chains. A **food chain** describes the relationship of who eats whom in an ecosystem. Here is the typical five-step order of a food chain with an example:

1. **Producer**—captures the energy from a source (typically sunlight) and makes food (e.g., grass)

2. **Primary consumer**—eats the producer (e.g., grasshoppers, beetles)

3. **Secondary consumer**—eats the primary consumer (e.g., sparrows, robins, bluebirds)

4. **Tertiary consumer**—eats the secondary consumer (e.g. hawks, owls, eagles)

5. **Decomposer**—breaks down all dead producers and consumers, and releases their nutrients back into the ecosystem (e.g., bacteria, snails, fungi)

Each order or eating level in the food chain is called a **trophic level**. Producers are the lowest trophic levels, while tertiary consumers, or beyond, are the highest; food chains rarely have more than four trophic levels. Decomposers process food from all trophic levels. Because most consumers eat two or more types of organisms that are often at different trophic levels, several food chains often combine to form a complex network called a **food web**. A consumer that feeds at multiple trophic levels (typically both plants and animals) is known as an **omnivore**, and these are present in almost every food web.

AP EXPERT TIP

Be careful when determining trophic levels. For example, the *secondary* consumers are actually on the *third* trophic level.

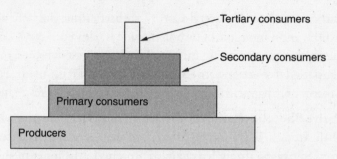

Figure 4.4 Hypothetical Ecological Pyramid

ECOLOGICAL PYRAMIDS

If you count the numbers of individuals at each trophic level in a food chain or food web, there tend to be more producers than primary consumers, more primary consumers than secondary consumers, more secondary consumers than tertiary consumers, and so on. If you graph the number of organisms at each trophic level, the graph will form the shape of a pyramid, called a **pyramid of numbers**. (See Figure 4.4.) The same pyramid relationship holds up when you plot the total mass of organisms at each level (i.e., **pyramid of biomass**).

The pyramid relationship is a direct result of the way in which energy is transferred from one trophic level to another. Remember that the process of respiration is not 100 percent efficient, and some energy gets lost as heat at each trophic level. The size of each trophic level in the pyramid represents the relative numbers, biomass, or energy available. To compensate for this energy loss, individuals at each trophic level must eat more individuals of the trophic level below it to maintain energy. So a graph of the energy available to each trophic level would also be pyramid shaped (i.e., **pyramid of energy**).

With this in mind, let's look at the following features of energy flow through an ecosystem (see Figure 4.5):

- Energy flow through an ecosystem moves in one direction from the energy source to the producers through the consumers.

- Energy becomes decreasingly available as trophic level increases.

- Some energy gets lost as heat at each transfer from one trophic level to another.

In contrast to energy, all matter in the ecosystem gets recycled. (See Figure 4.5.) Matter gets incorporated into producers and is subsequently passed on to consumers in higher trophic levels. When producers and consumers die, the decomposers

AP EXPERT TIP

This is typically referred to as the "10 percent rule," meaning that only about 10 percent of the energy from one trophic level is actually passed on to the next trophic level.

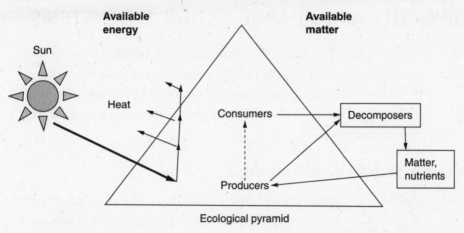

Figure 4.5 Flow of Energy and Matter Through an Ecosystem

recycle their matter back into the ecosystem. (Some matter may be transferred from one ecosystem to another during weathering processes.) Because the trophic levels have a pyramid structure, there are two points about how matter proceeds through the ecosystem:

1. When new matter (such as a herbicide) is introduced into the ecosystem, it appears in an orderly fashion over time (i.e., producers, then primary consumers, then secondary consumers, etc.).

2. As the new matter passes through the food chain, it gets more concentrated in organisms of each higher trophic level. This increase is called **biomagnification**.

For example, in the 1950s through the 1970s, the pesticide dichloro-dyphenyl-trichlor-ethane (DDT) was widely used on crops (producers). Although the concentration of pesticides in each individual crop plant was low, primary consumers (mice and hares) each ate many individual plants, absorbing and retaining a little DDT from each, and ultimately concentrating the pesticide in their bodies. Eventually, eagles (tertiary consumers) accumulated high concentrations of DDT from eating many lower level consumers. In eagles and other birds of prey, DDT interfered with eggshell formation and almost drove the bald eagle to extinction. The use of DDT was subsequently banned in the United States. Although new DDT has not been added to the environment, the existing DDT will remain until it ultimately breaks down chemically (a process that takes decades or longer) or gets cleaned up by humans (i.e., remediated from the soil).

AP EXPERT TIP

DDT became more concentrated at the top of the food chain because there is less biomass at this level, so the toxin is spread out among less mass of individuals.

DIVERSITY AND CHANGE WITHIN ECOSYSTEMS

Communities in ecosystems can be diverse in that they have large numbers of different species and that these species are evenly distributed throughout the ecosystem. But how does an ecosystem become diverse? The primary mechanism leading to diversity of species is natural selection and subsequent evolution. Let's look at these mechanisms.

NATURAL SELECTION AND EVOLUTION

Charles Darwin formulated a theory of evolution that could explain not only that many species could evolve from one but also how this evolution could take place. He published his ideas in a book entitled *The Origin of Species*. Darwin's theory of evolution has the following points to it:

- Organisms reproduce similar organisms: A dog reproduces a dog, a dandelion reproduces a dandelion, and a fish reproduces a fish.

- Often, the number of offspring is overproduced: The number that survive is less than that initially reproduced.

- Individual organisms must compete with each other and with other species for limited resources in the environment (food, water, space, etc.).

- In any population, individuals vary with respect to any given trait (e.g., height, skin color, fur color, shape of beak), and these variations can be passed on to the next generation.

- Some variations are favorable (i.e., make those individuals better suited to their environment) and some are not.

- Those organisms with favorable variations will survive and pass those traits on to their offspring, while those individuals with unfavorable variations will die and not pass on their traits (**natural selection**).

- Given sufficient time, because natural selection will accumulate these favorable traits, the species will change or evolve.

While there were other theories about evolution at the time, what made Darwin's theory unique was the idea of natural selection. He had known that mankind was artificially selecting animals and plants for thousands of years to produce new breeds of plants (corn, rice, wheat) and animals (dogs, cats, cattle) through selective breeding. Why couldn't nature do the same thing operating on a time scale of millions of years? For example, here are seven steps outlining how Darwin's theory of evolution explained the giraffe:

AP EXPERT TIP

Natural selection is often referred to as "Survival of the Fittest," where "fittest" is used in the sense of "those most suited for their environment," not necessarily those that were in the best physical condition, as is commonly used today.

1. Start with the short-necked ancestral giraffes.

2. With each generation, there were variations among individuals according to neck length: Some had short necks and some had slightly longer necks.

3. Those individuals with longer necks could reach both high and low leaves on the trees for food, whereas the shorter-necked individuals could only reach low-level leaves.

4. In times of drought, the slightly longer-necked individuals could reach higher vegetation and survive, but the short-necked ones would die off (i.e., natural selection).

5. The slightly longer-necked giraffes would pass their trait on to the next generation.

6. In the next generation, there would be variations again, but the average neck length might be slightly longer than the average neck length of the previous generation.

7. After millions of years, the average length of the giraffe's neck would be much longer than the ancestral form.

This process of natural selection takes such a long time because not every generation is exposed to some condition that selected for the long neck; if there were no droughts during a particular generation, then the short-necked variations might survive equally well as the long-necked variations. You must note that evolution works through variations within populations and that natural selection works on these variations to favor those organisms best adapted to the environment.

Over time, natural selection can change species and even produce new ones. If we consider the variations of a trait in a population, we will see that most individuals will have one form of variation (average) and a few individuals will have other forms (extremes). In other words, the frequency of variation within a population will initially follow a normal distribution, or "bell-shaped" curve. Now we can see the following ways that natural selection can affect the population:

- **Directional selection**—an environmental change occurs that gives an advantage to a variation on one end of the distribution and against the other (e.g., longer-necked giraffes over shorter-necked giraffes). Over time, the entire distribution shifts toward that end.

- **Stabilizing selection**—selection for the average form and against the extremes. Over time, the shape of the population curve becomes narrow at the mean and the mean increases. For example, perhaps snails living on a rocky shore have different colored shells ranging from white to black with gray being the average. The gray shells blend in with the rocks best, while birds easily recognize and eat the white-shelled and black-shelled snails. The gray-shelled snails survive and pass those traits on to their offspring. Over time, the frequency of gray-shelled snails in the population increases.

- **Disruptive selection**—selection for the extremes and against the average. Over time, the distribution divides into two separate populations (one for each extreme). Consider a rocky mountainside with patches of snow and black rocks. On this mountainside lives a population of mice whose fur color varies from white to black with average being gray. Hawks and other predators can pick out the gray mice easily and eat them, but the white or black ones can more easily hide against the snow and rocks. The white and black mice survive and pass these traits on to their offspring. Over time, one population (mainly gray mice) gives rise to two populations and eventually two species of mice (white and black).

A population can get split into two or more subpopulations by isolation. Natural selection then works on the isolated subpopulations to produce new species. Isolation can come in the following forms:

- **Geographic isolation**—some geographic barrier (e.g., mountain range, ocean, road) separates a population. Over time, the parent population and isolated population become more and more different as the dominant traits present in each population at the time of separation become more and more prevalent. This may also occur if individuals of one population end up in a new location and are subjected to different environmental circumstances, such as when birds blown off course land on an isolated island.

- **Reproductive isolation**—members of a population cannot reproduce freely with other members:

  — **Time or temporal**—males and females of the same population do not mature sexually at the same time to breed with each other, but they might breed with members of another population. Temporal isolation is common in plants when the male parts ripen at a different time than the female parts (this also prevents self-pollination).

  — **Mechanical**—the reproductive organs do not fit well enough to work between some members of the population, but work well with others. This creates two subpopulations.

  — **Behavioral**—courtship rituals are only recognized by some females of a population and not by others.

  — **Gamete mortality**—the sperm and egg of members within a population (and between different species) may not be compatible or their union may not produce viable offspring.

- **Ecological**—members of a population may occupy different sections within a habitat or ecosystem.

The mechanisms of natural selection and evolution are powerful and allow species to diversify. These isolation mechanisms may start out as occasional occurrences, but become more common over time, reinforcing separation between populations and giving rise to new species. These same mechanisms then serve to prevent mixing (hybridization) between different existing species.

ECOSYSTEM SERVICES

The diversity of species within ecosystems and the diversity of ecosystems themselves leads to the question "What do ecosystems do for people?" Ecosystems provide resources or functions for humans called **ecosystem services**. Ecosystem services can be categorized into the following:

- **Products**—timber, food (e.g., fisheries), medicinal plants
- **Functions**—waste decomposition and detoxification, air and water purification, flood control, soil conservation, and erosion control

Let's look at examples of ecosystem services. The temperate and boreal forests of North America are rich in hardwood trees (e.g., oak, maple, ash, hickory). The trees provide timber that is used in several industries including construction and furniture making. These industries contribute billions of dollars to the U.S. economy.

Wetlands along rivers such as the Mississippi River provided sinks for overflow water whenever the river flooded its banks. However, draining these wetlands for agriculture and housing development and the subsequent constructions of levees and dams interfered with the natural movement of the river and its channels. In the early 1990s, excessive runoff caused the Mississippi to swell, break levees, and spill out into farmlands and towns along the river, thereby causing huge damage to property in the billions of dollars. Thus, the wetlands had provided an important flood-control service.

In addition to flood control, wetlands also purify water and do so in a much more efficient and inexpensive manner than human technology. In New York, destruction of wetlands (for housing development) along the Catskill Mountains drainage basin led to a loss of water quality in New York City. City officials found that it was more economical to buy up property and restore the wetlands than to build new water-treatment facilities.

Islands along the Atlantic Coast of the United States provide an important barrier between the mainland and the ocean. The sands of these barrier islands shift with currents and storm surges caused by hurricanes and minimize damage to the mainland. However, because the barrier islands are popular vacation spots, they are being developed with houses and condominiums. Seawalls have often been constructed to prevent beach erosion. These activities limit the natural mobility of the island sands. When hurricanes do impact the islands, they often change the shapes of the beaches and destroy beachfront property.

In summary, ecosystem services, and their possible loss, should be considered in every aspect of land management. Ecological impact studies are a good tool to use prior to any development projects.

Ecosystem Changes

Individuals, populations, species, communities, ecosystems, and the entire biosphere are not static. Instead, they are subjected to constant environmental changes due to natural causes and human activities. Some environmental changes (e.g., climate shifts) can occur over a long time (years, decades, centuries, etc.), while others (geologic, pollution) can happen rapidly (hours to days). Let's examine some types of environmental changes.

Climate Shifts

As mentioned in Chapter 3, climate is the average weather in a given area over a long period of time (at least 30 years). Climate takes into account general weather patterns, seasonal changes, temperature, and precipitation. The two most important factors in climate are temperature and precipitation. Climates can be affected by the following:

- Global air-circulation patterns
- Ocean currents
- Atmospheric composition

A pair of vertical, opposite-moving convection cells occurs for every 30° change in latitude. The rising moist air and rainfall support boreal forests, and the falling cold, dry air supports deserts (cold and hot). Small shifts in these patterns would cause shifts in vegetation and changes in biomes (dryer airs would lead to grasslands, shrublands, and deserts, while moist air would lead to deciduous and coniferous forests).

The difference in heating between seawater at the equator and seawater at the poles sets up flows, or **currents**, of warm water. These currents help mix the waters of the world's oceans and influence weather patterns. For example, the Gulf Stream brings warm water to Iceland and Great Britain; without it, these places would be much colder.

Another example of how ocean currents influence climate periodically is the *El Niño*-Southern Oscillation (ENSO) that occurs in the Pacific Ocean every few years. The four-step sequence of events in the ENSO is as follows:

1. Trade winds blow westward and move warm surface waters toward Australia. This allows cold water, which is rich in oxygen and nutrients, to rise to the surface in the Pacific Ocean off the coast of South America. This cold water is important for fish and other sea animals and plants.

2. During *El Niño*, the trade winds weaken and warm water spreads across the Pacific Ocean. Because these warm waters have little oxygen and nutrients, sea life cannot live in them. *El Niños* are followed by *La Niñas*.

3. During a *La Niña*, strong trade winds blow warm surface waters west and allow cold waters to spread across much of the Pacific Ocean.

4. Because warm surface waters cause heavy rains, *El Niños* and *La Niñas* can change weather patterns in Australia, Southeast Asia, and even North America. *El Niños* cause droughts in Australia and Southeast Asia and heavy rainfall in North America. The opposite is true during a *La Niña* event.

The ENSO cycle occurs in irregular two-to-seven-year intervals, and an event can last for 12–24 months. Again, the changes in rainfall cause climate changes that influence vegetation (moist conditions producing deciduous, coniferous trees, dry conditions producing grasslands, shrublands, deserts).

One controversial influence on climate is **global warming**. Greenhouse gases (carbon dioxide, methane, water vapor) in the atmosphere can trap and hold heat from the Sun. In addition, human pollution (burning fossil fuels) adds more greenhouse gases to the atmosphere. The overall effect is that average global temperatures are on the rise. A small change (even a 5°C increase) could be enough to cause polar ice sheets to melt, sea levels to rise, and shifts in global air-circulation patterns northward. These climate shifts would cause changes in the distribution of biomes. The ENSO cycle is expected to speed up as a result of global warming, resulting in *El Niño* and *La Niña* events of greater strength and frequency. Scientists and governments disagree as to how to approach global warming and its implications.

GEOLOGIC CHANGES

Geologic changes such as drifting continents can cause changes in climates. For example, the movement of the African continental plate northward has shifted and expanded the Sahara Desert. Geologic evidence (fossils, rocks) shows that sub-Saharan Africa was once much wetter than it is now. The formation of the Great Rift Valley in East Africa changed the climate from tropical rain forests to grasslands and savanna.

Geological changes can cause more rapid ecological changes. For example, volcanic eruptions can destroy whole ecosystems in a matter of hours to days. In 1980, Mt. St. Helens in Washington State erupted and the ash and mudflows destroyed whole boreal forests. Twenty years later, life is slowly returning to that area, but it has not completely recovered to its pre-eruption state.

OTHER CATASTROPHIC CHANGES

Besides geology, fire and floods can cause rapid changes to ecosystems. Human pollution can cause changes to ecosystems slowly or rapidly. For example, an oil supertanker called the *Exxon Valdez* broke apart in Prince William Sound in Alaska. The oil spilled over a vast area of the Sound and disrupted aquatic and coastal ecosystems.

MOVEMENTS OF SPECIES

So species, communities, and ecosystems must be able to adapt over time to environmental challenges (also called environmental stresses or pressures). Natural selection and evolution will ensure that those organisms that are best adapted to the change will survive, while ill-adapted organisms will die out. But natural selection and evolution take a long period of time. For many organisms, populations, and species, moving away might be the best way to deal with an environmental stress. Let's examine how organisms and populations can change by moving.

When you look at an area occupied by a given ecosystem, not all populations are evenly distributed throughout. Populations can also be randomly distributed or grouped into clusters. For example, Americans tend to be clustered along the coastal states of the United States and in a few heartland cities. So, many populations can move or **migrate** either within the ecosystem or to another ecosystem. Migration can be in two forms:

1. **Active**—Individual organisms can walk, swim, run, or fly within and outside of the ecosystem. For example, Canada geese can fly within their ecosystem or migrate across North America depending upon the season of the year.

2. **Passive**—Moving air (winds, storms), water (runoff, currents, tides), or other organisms (birds, insects) can carry an individual or its offspring within or out of the ecosystem. For example, many trees rely on wind to disperse seeds.

Migration within an ecosystem may be a response to a daily change. For example, microscopic marine animals (zooplankton) usually migrate to deeper regions of the ocean during the day and return to the surface at night, presumably in response to changes in light.

Alternatively, migration within the ecosystem may be seasonal. For example, mule deer in Southern California spend summers grazing on the cool north-facing mountain slopes and spend winter on the south-facing slopes. We are most familiar with round-trip migrations between ecosystems that occur seasonally. The flyways of North America are filled with migrating birds each autumn and spring. The birds move from Northern deciduous and boreal forests in the summer to spend winter in the tropical/subtropical deciduous and rain forests in the South. In the winter, caribou migrate from summer feeding grounds in the Taiga to the Arctic Tundra, where they feed on lichens. Most of these migrations are stimulated by seasonal changes in the ecosystems.

A second type of migration involves a one-way trip. Pacific salmon hatch in the freshwater rivers and streams of the Pacific Northwest, but they swim to the Pacific Ocean where they live their lives. When they become sexually mature, they migrate back to the very rivers and streams where they were hatched. There, they mate and die.

A third type of migration is a multigeneration round-trip. The monarch butterfly spends its summers in Northern America. In the fall, the butterflies migrate southward into Mexico.

In January, they begin a northward movement into the southern United States, where they mate and die. Subsequent generations develop and work their way northward into the northern North American continent. By the time the Monarch butterflies return to the northern summer grounds, they are several generations removed from those that left in the fall.

Migrations not only allow populations to respond to environmental challenges in their native ecosystems, but also allow them to infiltrate and colonize new ecosystems. So, migrating birds and insects can occupy niches in diverse ecosystems seasonally (e.g., deciduous and boreal forests in the summer and tropical forests in the winter).

ECOLOGICAL SUCCESSION

Many individual organisms, species, and populations may respond to environmental challenges by natural selection or evolution and/or by migration. But how does an entire ecosystem respond to an environmental stress (climate change, fire, pollution, etc.)? Ecosystems respond by a process called **succession**. Succession is a series of replacements, one community by another, until a stable community called a **climax community** is reached. Following are the two types of succession:

1. **Primary**—New ground devoid of life (such as volcanic rock) is colonized, and a new ecosystem develops through the process of succession.

2. **Secondary**—Part of an existing ecosystem is disturbed by some event (fire, climate change, volcanic eruption), and the ecosystem responds through succession, eventually restoring the disturbed area. In these areas, soil is already present.

Regardless of whether the succession is primary or secondary, the four phases are similar:

1. **New land is exposed**—The new land could be devoid of life (primary succession) or could be a previous stage caused by some disturbance in an existing ecosystem.

2. **Pioneer species take root**—These organisms, usually vegetation, are fast-growing, short-lived plants that can live in exposed conditions (drastic changes in temperature, plenty of sunlight, wind, loose or rocky soil) depending upon the ecosystem. These species could be lichens or mosses. The pioneer species begin to modify the ecosystem (stabilize loose sand or soil, retain moisture, etc.) and make way for other species.

3. **Successive seral stages become established**—Longer-lived plants become established (perennial weeds, grasses, herbaceous plants). They can either replace or coexist with the pioneer species. Several such stages occur as each

AP EXPERT TIP

Primary succession occurs in an area where there are no remnants from a previous ecosystem.

AP EXPERT TIP

Pioneer species are usually shade intolerant species, like grasses.

stage modifies the ecosystem further (modifying soil, retaining water, providing nutrients from decaying matter, reducing wind, providing shade, etc.).

4. **Subclimax seral (intermediate) stages become established**—Longer-lived, taller plants and trees become established (shrubs, young hardwood trees, and conifer trees).

5. **Climax community gets established and remains**—The "end" of succession occurs when a stable community exists, and it can be categorized as a "mature" forest with older vegetation. The ecosystem remains in this form until a disturbance alters it and secondary succession begins. Some ecologists, though, feel that climax communities can never be reached, since ecosystems are always changing (e.g., a tree falls down and creates an opening where the trees in the undergrowth can prosper and dominate).

As the vegetation in each phase of succession becomes established, animals colonize the community (insects, birds, reptiles, mammals, etc.) and establish themselves and species interact (competition, predation, symbiosis), thereby making the community distinct.

Let's look at a seven-step example of primary succession, where bare rock becomes exposed as a glacier recedes:

1. Lichens invade and establish footholds on the exposed rocks. The lichens help to retain moisture. When they die, their decomposed remains mix with the bits of weathered rocks, provide nutrients, and form soil. (Pioneer species take root.)

2. Mosses invade and mix with the lichens. Mosses attract insects and other animals. Acids from the mosses break down the rocks. The rock pieces combine with decayed remains of mosses and animals to further enrich the soil. The mosses outcompete and replace the lichens. (A seral stage becomes established.)

3. Grasses and weeds can now grow in the richer soil and attract other animals (insects such as grasshoppers and beetles). (Successive seral stages become established.)

4. Herbaceous plants (goldenrod, milkweed, ragweed, wild carrots, wild onions) can grow in the rich soil. Small animals (snails, moles, mice, voles, rabbits) invade the community and graze on plants. The plants retain moisture and nutrients in the soil, and the decay of dead plants and animals further enriches the soil. (Successive seral stages become established.)

5. Shrubs and taller, woody plants (sumac, wild rose, mulberry) can grow in the rich soil. Deer can graze on the plants and shrubs. Predators such as the red fox can prey on mice and voles. (Subclimax seral stages become established.)

6. Sapling trees and wildflowers invade and establish themselves in the shrub community. They can alter the conditions of wind, light, and temperature. They provide enough organic matter to keep the soil rich and attract more animals (squirrels, raccoons, birds) to the community. (Subclimax seral stages become established.)

7. Large hardwood trees (maple, oak, hickory) can exist in the rich, moist soil and form a fully mature forest community. (Climax community is established.)

This example of succession takes thousands of years to occur. In contrast, in rapidly changing environments such as sand dunes, succession can occur quickly in a matter of months or years and lead to different climax communities (grasses and shrubs).

Succession can also occur in aquatic ecosystems, such as a small pond or shallow lake (primary succession). Plankton become the pioneer species. They may attract various fish (blue gill, bass) from the watershed of the pond or lake. As the pond or lake accumulates sediments from the watershed, marsh grasses become established. The grasses retain more sediment and the pond or lake becomes shallower. As new species of plants invade and establish themselves, they absorb water, retain more sediment, and alter the mud into soil. Eventually, the pond or lake dries up, and the aquatic community succeeds into a land community.

BIOGEOCHEMICAL CYCLES

Energy flow through an ecosystem is a one-way movement through living organisms. In contrast, as matter flows through an ecosystem, it gets recycled. Many chemicals are important for the life cycle between living organisms, the atmosphere, the oceans, and Earth's crust. These chemicals include carbon, oxygen, nitrogen, phosphorus, sulfur, and water. The cycles of these chemicals are called **biogeochemical** cycles (bio: life; geo: rocks, air, water; chemical: chemical interactions). Biogeochemical cycles involve multiple ecosystems, consist of many processes, and have global effects.

THE CARBON CYCLE

Carbon is the main element in all living organisms. Besides being found in living organisms, carbon can be found in the atmosphere as carbon dioxide and in the oceans and rocks as carbonates. The carbon cycle is the exchange of carbon, mostly as carbon dioxide or carbonates, between the oceans, the atmosphere, and the rocks. (See Figure 4.6.)

The carbon cycle involves at least five pathways:

1. Carbon dioxide moves through living organisms:

   - Gaseous carbon dioxide gets fixed into plants and microorganisms (producers) through photosynthesis.

   - Carbon passes through the food chains and webs as consumers eat the producers and each other.

   - Fixed carbon in food (glucose, fats, proteins), wastes, and dead organisms get broken down into carbon dioxide and released into the atmosphere by respiration in producers, consumers, and decomposers.

   - Carbon from decomposition makes its way into the soil. From the soil, it can be taken up into plants, or over millions of years it can become trapped and compressed into rocks (e.g., fossil fuels: coal, oil, natural gas).

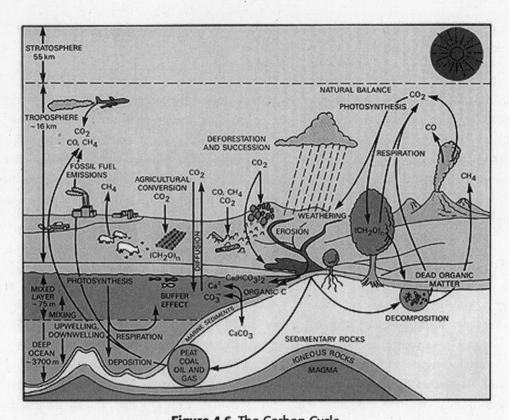

Figure 4.6 The Carbon Cycle

Source: Courtesy of NASA, available at www.rst.gsfc.nasa.gov.

2. Carbon dioxide from the atmosphere moves into the oceans:

   • Carbon dioxide from the atmosphere dissolves in the oceans and combines with water to from carbonic acid and carbonates.

   • Dissolved carbonates in the ocean get incorporated into the shells of microorganisms (e.g., diatoms, plankton, marine invertebrates). When these organisms die, the carbonates settle to the bottom ocean sediments.

3. Carbonates form rocks:

   • Over time, sediments containing carbonates get compressed to form sedimentary rocks containing carbonates (limestone).

4. Geological forces return carbon from the rocks to the atmosphere:

   • Plate tectonic movements cause ocean sediments and oceanic crust to pass underneath continental crust at subduction zones.

   • The heat and pressure melts the rocks, releasing their carbon dioxide that rises to the surface and spews into the atmosphere from volcanoes.

   • Erosion of exposed carbonate rocks (weathering) causes carbonates to return to the sea.

5. Human activities emit carbon dioxide into the atmosphere:

   • Burning fossil fuels release carbon dioxide.

   • Farm and domestic farm animals release carbon dioxide through the digestion (i.e., oxidation) of carbon in food.

It is difficult to say precisely how long a molecule of carbon dioxide remains in the atmosphere (residence time), as there are different ways of measuring this. However, Intergovernmental Panel on Climate Change estimates range from approximately 100 to as many as 400 years. Another carbon-based molecule, methane (CH_4), has a residence time of about 12 years.

If you look at the distribution of carbon between the atmosphere, the oceans, and the rocks, you will find that most of Earth's carbon is trapped in the rocks and little is in the atmosphere. Carbon dioxide in the atmosphere, along with methane and water vapor, tends to trap heat from the Sun. So, the amount of carbon dioxide in the atmosphere plays an important role in regulating Earth's surface temperature. If there is too much carbon dioxide, global temperatures will rise (greenhouse effect and subsequent global warming). Current debate on this subject involves how much carbon dioxide is added to the atmosphere by burning fossil fuels and whether increasing global temperature is directly related to burning fossil fuels. (The scientific consensus is that it is.)

THE NITROGEN CYCLE

Nitrogen is another biologically important element that is part of all proteins and nucleic acids. The major source of nitrogen is Earth's atmosphere, which is 78 percent nitrogen gas. However, nitrogen gas is not available to living organisms. Nitrogen gas must first be fixed to a chemically usable form such as nitrates or nitrites. (See Figure 4.7.)

Let's look at five important steps in the nitrogen cycle:

1. Nitrogen gas must be made into a chemically usable form (**fixation**). Some species of nitrogen-fixing bacteria have a mutualistic relationship with legumes. They reside in the roots of legume plants where they convert nitrogen gas from the atmosphere into ammonia in the soil. (Other species of nitrogen-fixing bacteria also live in the soil and perform the same functions.)

2. Nitrifying bacteria in the soil convert ammonia to nitrites and nitrates (**nitrification**).

3. Now that the nitrogen is in a usable form, plants take up the nitrates from the soil and incorporate them into amino acids in their tissues. Animals eat the plants and incorporate the ingested amino acids into their own tissue (i.e., amino acids, proteins, nucleic acids). These steps are called **assimilation**.

4. When animals and plants die, decomposers convert the amino acids back into ammonia and return it to the soil (**ammonification** or **mineralization**). Nitrifying bacteria again convert the ammonia to nitrates and nitrites.

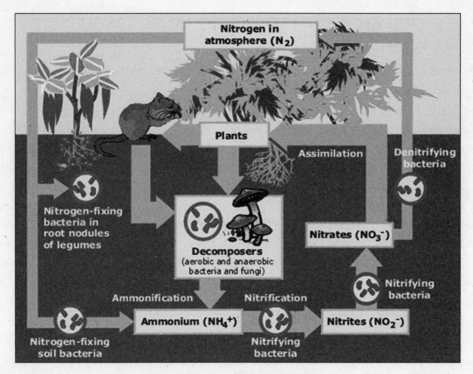

Figure 4.7 The Nitrogen Cycle
Source: Courtesy of USEPA.

5. Denitrifying bacteria in the soil convert nitrates back into gaseous nitrogen (**denitrification**), which escapes into the atmosphere, thereby completing the nitrogen cycle.

Nitrogen cycles in aquatic systems as well. In those ecosystems, cyanobacteria perform the nitrogen-fixation steps and other bacteria perform ammonification and denitrification steps. Furthermore, aquatic systems lose small amounts of nitrogen to marine sediments (which eventually get incorporated into rocks), but they gain some nitrogen when weathering washes nitrate-containing rock particles into streams and rivers.

Human activity can upset the balance of nitrogen cycling in the ecosystem, either by removing nitrogen from ecosystems or adding excess nitrogen. Destroying forests and plants removes organic nitrogen sources, thereby resulting in a steady decline of nitrogen in the soil. In contrast, commercial fertilizers for agriculture add too many nitrates to the ecosystem. These nitrates can also run off into streams and rivers, thereby causing disturbances in nitrogen cycling of aquatic ecosystems. Similarly, the discharge of human wastes and untreated sewage into rivers, streams, and estuaries can add further nitrogen loads to aquatic ecosystems. Finally, automobile and power plants emit nitrogen dioxide into the atmosphere. The nitrogen dioxide reacts with oxygen to form ozone that remains in the lower levels of the atmosphere. (This reaction is especially abundant on hot summer days.) The ozone is highly reactive and detrimental to living organisms. (In humans, it exacerbates respiratory conditions such as asthma and emphysema.)

THE PHOSPHORUS CYCLE

Phosphorus is also a biologically important compound. Many enzymes require phosphate for activation or inactivation. Phosphorus-containing ATP and adenosine diphosphate (ADP) are important for storing and using energy from photosynthesis and respiration. Unlike the carbon and nitrogen cycles, no pool of phosphorus exists in the atmosphere, so the phosphorus cycle is limited mainly to the soil and water. The major form of phosphate is the mineral apatite, found in rocks and phosphate deposits. (See Figure 4.8.)

Let's look at the four phases of the phosphate cycle in a terrestrial ecosystem:

1. Weathering of phosphate rocks leaches phosphate into the soil.

2. Like the carbon and nitrogen cycles, plants take up phosphate from the soil and incorporate it into their tissues. Likewise, animals eat the plants and assimilate phosphate into their tissues.

3. When the plants and animals die, decomposers release the phosphate back into the soil.

4. Animal excretions also contain phosphate that gets back into the soil.

Weathering and erosion leach some phosphate from the rocks and soils into streams and rivers. (See Figure 4.9.) Here, the phosphorus cycle is similar to

> **AP EXPERT TIP**
>
> Remember that phosphorus does not have a gas phase, and therefore does not exist in the atmosphere. On the exam, you may see an answer that refers to phosphorus existing in the atmosphere. Now that you know it's not true, you can immediately eliminate this answer.

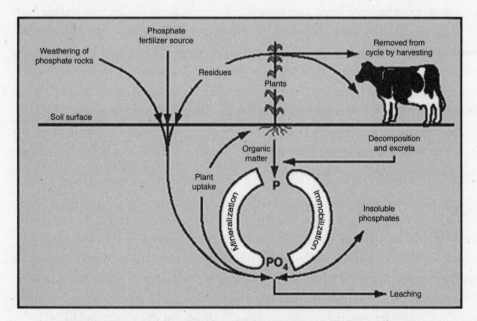

Figure 4.8 The Phosphorus Cycle in a Terrestrial Ecosystem
Source: Courtesy of USEPA, available at www.epa.gov/owow/watershed/wacademy/sam/
images/09_fig06.gif.

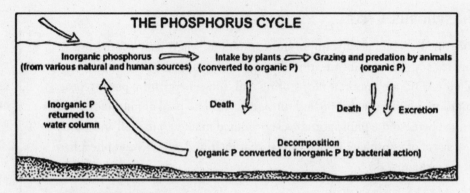

Figure 4.9 The Phosphorus Cycle in an Aquatic Ecosystem

Source: Courtesy of USEPA, available at www.epa.gov/owow/monitoring/volunteer/stream/images/ fig512.jpg.

terrestrial systems. Aquatic plants and phytoplankton assimilate phosphate. Marine vertebrates and invertebrates assimilate phosphate by eating plants, phytoplankton, and each other. When aquatic plants and animals die, bacterial decomposers break down organic phosphate to inorganic phosphate, which returns to the water. Excreted phosphate also returns to the water column. Finally, the water column loses some phosphate to marine sediments, which eventually get converted to phosphate-containing rocks by geological processes.

Human activities can overload the phosphorus cycle, like the nitrogen cycle. Excessive use of fertilizers used in agriculture places phosphorus loads on the terrestrial phosphorus cycle. These phosphorus loads can wash into rivers and streams, where, along with waste and sewage discharge, they can overload aquatic phosphorus cycles. Excess phosphorus in an aquatic ecosystem (**eutrophication**) can stimulate excessive growth of algae (algal blooms). The algal blooms can form thick mats on the water surface, but these algae are short-lived and die quickly, resulting in excessive dead organic matter in the water. The decomposition of the dead algae uses up the available dissolved oxygen (DO) in the water. The lack of oxygen in the water is a condition called **hypoxia**, which suffocates fish and other aquatic life. A water body is considered hypoxic when DO drops below 2 parts per million (the standard unit for DO measurement). Massive fish kills disrupt the food chains and webs of aquatic ecosystems by removing top-level consumers and proliferating decomposers.

THE SULFUR CYCLE

Sulfur is important for proteins, which have the sulfur-containing amino acids cysteine and cystine. The sulfur cycle (Figure 4.10) is similar to the phosphorus cycle. Sulfur is mainly found in rocks and soil (coal, oil, peat) as sulfate minerals. Weathering exposes sulfates from the rocks into the soil and into aquatic ecosystems. In both these ecosystems, plants and other photosynthetic organisms take up and assimilate sulfates into their tissues. Animals eat the plants and likewise assimilate sulfates into their tissues. Death and decomposition convert organic sulfates into inorganic sulfates. Animal excretions also add sulfates to the soil or water. Sulfates then recycle.

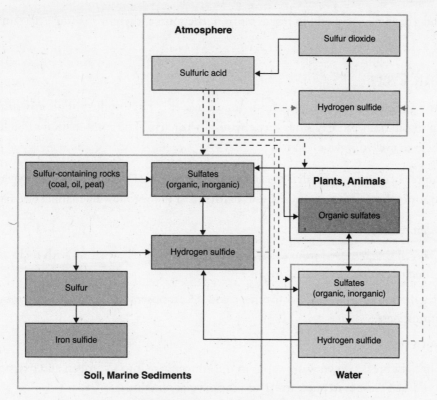

Figure 4.10 The Sulfur Cycle

However, unlike the phosphorus cycle, there is a component of the sulfur cycle that involves the atmosphere. During decomposition in both soil and water, decomposers convert sulfates into hydrogen sulfide gas that can escape into the atmosphere, the water, the soil, and marine sediments. Other major sources of atmospheric hydrogen sulfide include volcanoes and power plant emissions. Following are the possible fates of hydrogen sulfide depending upon where it is:

- In the soil, various chemosynthetic bacteria can convert hydrogen sulfide back into inorganic sulfates, to sulfuric acid, and/or to elemental sulfur. If iron is present in the soils or sediments, it will react with elemental sulfur to form iron sulfide, which gets incorporated into rocks by geological processes.

- In the water column, photosynthetic bacteria and other bacteria can convert hydrogen sulfide into organic and inorganic sulfates.

- In the atmosphere, hydrogen sulfide gas quickly breaks down into sulfur dioxide, where it combines with water vapor to form sulfuric acid. The sulfuric acid precipitates as acid rain, thereby returning sulfur to the soil and water columns. However, acid rain can also kill vegetation and erode rocks.

Emissions from coal-burning power plants dump enormous amounts of sulfur dioxide and sulfate particles into the atmosphere. Prevailing winds and storm systems carry the particles over vast distances and precipitate acid rain in places far from the source, thereby making acid rain a global problem.

THE OXYGEN CYCLE

Molecular oxygen is a critical substance for living things. It is a by-product of photosynthesis and a reactant in respiration. But oxygen is very chemically reactive. The combination of biology and chemistry help to cycle oxygen on Earth.

The main supply of oxygen on Earth is the atmosphere, which is 21 percent oxygen. Oxygen cycles between the atmosphere, living organisms, and Earth's crust in the following manner (see Figure 4.11):

1. Oxygen removal
   - Oxygen gets removed from the atmosphere by chemically reacting with rocks and minerals exposed by weathering.
   - Respiration by producers, consumers, and decomposers also removes oxygen from the atmosphere.

2. Oxygen supply
   - Sunlight breaks down water vapor in the atmosphere into hydrogen and oxygen. The hydrogen escapes to outer space and the oxygen remains.
   - Photosynthesis splits water to produce oxygen, which escapes into the atmosphere.

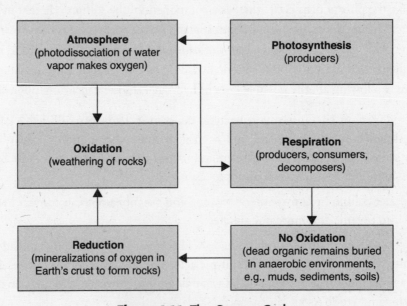

Figure 4.11 The Oxygen Cycle

Some oxygen in organic matter escapes oxidation when buried in muds and sediments (anaerobic environments). The remains get reduced and incorporated into rocks and minerals. These rocks and minerals eventually get weathered when exposed by geological processes.

THE WATER CYCLE

Water is important for all life on Earth. It is the solvent in which all biochemical reactions take place. Because of Earth's surface temperatures and pressures, water can exist in three states (solid, liquid, gas). Water cycles between the atmosphere, surface, and ground. Let's examine this cycle on Earth, which is called the **water cycle**, or **hydrologic cycle.** (See Figure 4.12.) Water is stored in the atmosphere (water vapor), on the surface (liquid water: rivers, streams, lakes, oceans; ice: glaciers, polar ice), and in the ground (liquid groundwater or ice in the form of permafrost). Water can move between these storage sites in six ways:

1. Absorption of the Sun's energy causes water to move from surfaces into the atmosphere (**evaporation**, **sublimation**).

2. Water from the ground and surface can move into the atmosphere through plants during photosynthesis in a process called **transpiration**. Therefore, scientists often combine the processes of evaporation and transpiration into one process called **evapotranspiration**.

3. Absorption of energy can melt snow and ice and transfer water to another storage site (lakes, oceans) on the surface or into the ground.

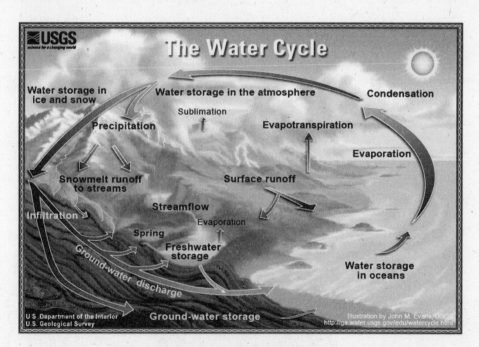

Figure 4.12 The Water Cycle

Source: Courtesy of USGS, Department of Interior (Illustration by John M. Evans and Howard Perlman), available at http://ga.water.usgs.gov/edu/watercycle.html.

4. Release of energy by water vapor in the atmosphere causes precipitation that returns water to the surface.

5. Once on the surface, water can move through tiny spaces in the soil (**infiltration**) and flow into groundwater storage.

6. Groundwater can also flow back to the surface and empty into streams, rivers, lakes, and oceans.

It is important to remember that solar energy powers the water cycle. Earth absorbs the solar energy and releases it back to outer space through the water cycle. The water cycle collects, purifies, and distributes Earth's water supplies. The water cycle helps establish weather patterns and storm systems and helps Earth deal with the solar radiation that it absorbs.

Human activities alter the water cycle in two ways. First, we can alter the freshwater supply by draining lakes, rivers, groundwater, and other reservoirs. This leads to infiltration by salt water into groundwater supplies. Second, by destroying vast amounts of forest, we reduce evapotranspiration and increase erosion, which reduces seepage into groundwater.

REVIEW QUESTIONS

1. Around hydrothermal vents in mid-oceanic ridges, there are species of tubeworms. In the tissues of these tubeworms are chemosynthetic bacteria. The bacteria provide the tubeworms with organic nutrients, and the tubeworms provide the bacteria with safe places to live and constant streams of mineral-rich water. Which of the following describes the relationship between the tubeworm and the chemosynthetic bacteria?

   (A) Predator-prey

   (B) Commensalism

   (C) Competition

   (D) Mutualism

   (E) Parasitism

2. European missionaries and traders who settled in the Hawaiian Islands in the 1800s imported the mongoose to eliminate snakes from homes, fields, and settlements. From the information given, which of the following BEST describes the mongoose?

   (A) Indicator species

   (B) Prey

   (C) Alien species

   (D) Native species

   (E) Keystone species

3. Elements cycle through the atmosphere in all of the following biogeochemical cycles EXCEPT which?

   (A) Carbon

   (B) Phosphorus

   (C) Water

   (D) Oxygen

   (E) Nitrogen

4. A battery manufacturing plant accidentally leached cadmium into a local lake. Ecologists sampled various species from the lake and determined the cadmium concentration in the tissues of each organism. The data are shown below:

	Cadmium Concentration (mg/kg tissue)
Species A	100
Species B	10
Species C	30
Species D	65
Species E	20

   Which species is most likely the second-order consumer?

   (A) Species A

   (B) Species B

   (C) Species C

   (D) Species D

   (E) Species E

5. Which of the following mechanisms of adaptation to environmental stress would NOT be applicable to a population?

   (A) Migration to another ecosystem

   (B) Natural selection

   (C) Succession

   (D) Migration within an ecosystem

   (E) Evolution

6. A change in the _____ cycle could induce a(n) _____ in an ecosystem.

   (A) oxygen, geologic change

   (B) water, acid rain

   (C) sulfur, eutrophication

   (D) nitrogen, hurricane

   (E) carbon, climate shift

7. Which of the following communities might be a stage in the primary succession of a coastal sand dune ecosystem?

    (A) Lichen communities supplant moss communities.

    (B) Beach grasses colonize the sands, thereby stabilizing the dunes.

    (C) Oak trees supplant the grasses on the dunes.

    (D) Shrubs supplant maple trees on the dunes.

    (E) Lichen communities establish themselves on the loose sandy soil.

8. Which of the following species is characteristic of a savanna?

    (A) Lichens

    (B) Oak trees

    (C) Douglas fir

    (D) Mangrove trees

    (E) Grass

9. The river that brings water to Townsville periodically floods and destroys homes upstream. To build the homes, wetlands were filled in. One proposal brought by local officials to fix the problem is to build a dam upstream of the homes and construct levees along the river to protect the homes and Townsville. Another proposal is to buy out the homes along the river and restore the wetlands. The second proposal is an example of using which of the following?

    (A) Biogeochemical cycles

    (B) Food chains and webs

    (C) Natural selection

    (D) Ecosystem services

    (E) Ecological succession

10. A company wants to place an electrical power line through a forested area. The company workers cut an area through the forest, erect the power lines and cables, and plant grass in the clear-cut area. All of the following are true about the clear-cut area EXCEPT which?

    (A) Clear-cutting is an ecosystem stressor.

    (B) The clear-cut area will induce an edge community.

    (C) The clear-cut area will initiate primary succession.

    (D) New species will migrate into the grasses of the clear-cut area.

    (E) The clear-cut area will initiate secondary succession.

FREE-RESPONSE QUESTION

A new aquatic ecosystem has been found in a small lake underneath a thick pack of ice in Antarctica. Measurements show that light can penetrate the ice into the water. Through drill holes, scientists have brought up samples and set up a laboratory on-site. The ecosystem appears to be entirely microscopic, and there are at least four new species of plankton in the water samples. You have been sent to investigate and characterize the new ecosystem.

(A) **Discuss** how you would determine that the energy source for the ecosystem is light.

(B) **Describe** the possible experiments that you might conduct to determine the trophic levels of each species (A–D), and **explain** how the results of those experiments might help you conclude which species is at which trophic level.

ANSWERS AND EXPLANATIONS

1. D

This question addresses your ability to recognize the various aspects of interactions between species. The bacteria reside within the tissues of the tubeworm, so this is clearly a symbiotic relationship, which rules out responses (A) and (C). In the relationship, the tubeworm benefits by getting organic nutrients from the bacteria, and the bacteria get a safe place and constant flow of water. In a commensalistic relationship, only one partner benefits, which rules out response (B). In a parasitic relationship, one partner is harmed, which rules out response (E). As both species benefit, the correct choice is mutualism, response (D).

2. C

This question addresses your ability to recognize the classifications of species within an ecosystem. From the information given, the mongoose did not indicate any change in the Hawaiian Island ecosystem, so it is not an indicator species and choice (A) does not apply. The mongoose hunts snakes, so it is a predator and not prey, which rules out choice (B). There is not enough information given to establish that the mongoose is a keystone species, which rules out choice (E). The mongoose had to be imported to the islands so it is not a native species, which rules out choice (D). Therefore, the mongoose is an alien species, and choice (C) is correct.

3. B

This question assesses your knowledge of the various biogeochemical cycles. The cycles of carbon, oxygen, water, and nitrogen all cycle through the atmosphere, which rules out choices (A), (C), (D), and (E). Only the phosphorus cycle operates exclusively through ground, water, and sediments without exchanging with the atmosphere. So choice (B) is correct.

4. C

This question assesses your ability to examine data about biomagnification. When a pollutant or isotope is introduced into an ecosystem, the highest order consumer will have the highest concentration of the substance in its tissues. Choice (A) is the highest order, in this example, and therefore, incorrect. Producers will have the lowest concentration, which rules out choice (B). Choices (D) and (E) are third-order and first-order consumers, respectively, and therefore, incorrect. Choice (C) has the third highest concentration of the pollutant, and it is therefore the second-order consumer, the correct answer.

5. C

This question assesses your knowledge of how various components of the ecosystem can adapt to environmental changes. Migrations are applicable to individual organisms, species, and populations, which rules out choices (A) and (D). Natural selection and evolution can work on species and populations, which rules out choices (B) and (E). Succession applies only to communities and ecosystems, not to populations, so choice (C) is the correct answer.

6. E

This question assesses your ability to recognize the types of environmental changes that would be caused by changes in a particular biogeochemical cycle. None of the biogeochemical cycles induce geological changes, but rather respond to them, so choice (A) is incorrect. Although the water cycle has rain in it, there is no disruption in the cycle that would lead to acidification of the rain, which does occur in the sulfur cycle; therefore, choice (B) is incorrect. Eutrophicaton or algal blooms are caused by disruptions in the phosphorus cycle, not the sulfur cycle, so choice (C) is incorrect. The nitrogen cycle has nothing to do with weather patterns or storms, so choice (D) is incorrect. Only the global levels of carbon in the atmosphere can alter the surface temperatures that could induce climate shifts, so choice (E) is correct.

7. B

This question assesses your understanding of succession and its application to a beach community. Lichens are a pioneering species that invade a rocky area and come before mosses, so choice (A) is incorrect. Oak trees are part of a subclimax, climax community and would supplant shrubs, rather than grasses, so choice (C) is incorrect. Shrubs are a subclimax community species and would not supplant maple trees (which would not be found on a dune anyway), so choice (D) is incorrect. Lichen communities establish on rocks, rather than loose sandy soils, so choice (E) is incorrect. Grasses are the pioneer species that prevent the dunes from shifting, so choice (B) is correct.

8. E

This question assesses your knowledge of the vegetation associated with various biomes. Lichens are associated with tundra, so choice (A) is incorrect. Oak trees are associated with temperate deciduous forests, so choice (B) is incorrect. Douglas fir trees are conifers in boreal forest, so choice (C) is incorrect. Mangrove trees are found in tropical coastlines, ruling out choice (D). Grasses are associated with semiarid savannas, so choice (E) is correct.

9. D

This question assesses your ability to recognize an ecosystem service (flood control). Biogeochemical cycles, food chains and webs, natural selection, and succession have nothing to do with flood control. So this rules out choices (A), (B), (C), and (E). Flood control is an ecosystem service of wetlands, so choice (D) is correct.

10. C

This question assesses your ability to apply knowledge of ecosystem changes. Clear-cutting is a stressor to the forest ecosystem, so choice (A) is incorrect. The clear-cut area will induce an edge between the newly planted grasses and the forest, so choice (B) is incorrect. New species will migrate into the grassy area as the edge community develops and succession proceeds, so choice (D) is incorrect. Because previous life existed and exists in the clear-cut area, this is an example of secondary succession, so choice (E) is incorrect; therefore, choice (C) is correct.

ANSWER TO FREE-RESPONSE QUESTION

Most ecosystems get their energy from the Sun, and this energy gets passed through the trophic levels from producers to consumers. You must come up with possible experiments to determine the source of energy for the newly discovered ecosystem and the trophic levels.

(A) The passage mentioned that light could permeate the ice into the lake, so you might hypothesize that the energy for the food chain could come from light, and some organisms in the plankton would be photosynthetic. You could test this by incubating two samples, one in light conditions and one in dark conditions, and observe them over time to see which sample lives and which one dies. If the producers in the ecosystem are photosynthetic, then the samples incubated in the dark will die out, while those incubated in the light will survive. If the basis of the food chain is some other type of energy, then the results of this experiment indicate that it is nonphotosynthetic, but it will be inconclusive as to the exact nature of the energy source.

(B) There are three possible ways to try to determine the trophic levels of the ecosystem:

1. You could observe the population over time in an attempt to determine who eats whom. And you would have direct observational evidence of the food chain.

2. You could count the numbers of each species in many samples and construct a pyramid of numbers. The largest number of organisms would indicate the producers and the smallest number would indicate the top-level consumers. Intermediate numbers would indicate which species belonged to first- and second-order consumers.

3. You could introduce small amounts of a radioactively labeled nutrient or foreign substance (e.g., heavy metal) into several sample containers. Take samples from each container over time, isolate each species, measure the amount of radioactivity or the concentration of foreign substance in the tissues of each species, and plot these values over time. When a pollutant or radioactively labeled nutrient is introduced into an ecosystem, it follows a pattern over time. It first appears in producers, then first-order consumers, second-order consumers, and so on, until the last stage is the top-level consumer. Also, the concentration of the label or substance increases in the tissues as the trophic level increases (lowest in producers, highest in top-level consumers). A graph of this type would indicate which trophic level was occupied by each species.

CHAPTER 5: DYNAMICS AND IMPACT OF POPULATION GROWTH

IF YOU LEARN ONLY FIVE THINGS IN THIS CHAPTER . . .

1. Populations, whether nonhuman or human, do not grow indefinitely. There are limits to population growth. In nonhuman populations, growth can be described by an equation that takes these limits into account: $\frac{dN}{dt} = rN\left(1 - \frac{K}{N}\right)$, where K is the carrying capacity or limits of the environment. K can be determined by density-dependent factors (competition, predation, disease) and density-independent factors (food supply, temperature, climate).

2. An important descriptive tool to characterize human populations is the age-structure diagram that gives information about the makeup of any human population and can be used to predict changes in a given population (male to female ratio, booms and busts, dependency loads). Age-structure diagrams have distinctive shapes for various types of population growth (rapidly growing, slowly growing, zero growth, negative growth).

3. When studying the dynamics of human population growth, politics, economics, religion, and sociology become more important factors than just biology. The status and roles of women, religious beliefs, and socioeconomic conditions largely determine birth rates and fertility rates of any given country. The levels of poverty and of health care largely determine death rates. Immigration laws regulate migration rates.

4. Strategies for regulating population size usually involve influencing birth and fertility rates through family planning and regulating immigration. Countries will try to match population growth to the availability of resources and a desired standard of living. Governments can set population policies either directly or indirectly. The type of government (democratic versus totalitarian), economics, and the social and religious beliefs of the population will largely determine the success or failure of these policies.

5. Human populations can have an enormous impact on habitats. Increasing human population growth places consumption demands and environmental stresses on habitats and ecosystems. The consequences of these demands and stresses can affect ecosystems, both locally and globally.

CONCEPTS IN POPULATION BIOLOGY

Life has flourished on Earth. In almost any place you can think of, there are living organisms (microorganisms or more complex life forms). These organisms grow and thrive. But can life grow indefinitely? If not, what limits the growth of any organism? How many organisms can Earth support? These are but a few of the questions addressed by the study of populations called **population biology**. Population biology is the branch of ecology and environmental science that deals with characterizing the makeup, growth, and impact of populations of organisms on the environment. Population biology consists of the following disciplines:

- **Population dynamics**—Concerns the growth and limitations of the population and how that population interacts with its environment with respect to growth and stability.

- **Demography**—Describes the vital statistics of the population. What is its makeup (males versus females, age structure, population density, birth rates, death rates, migration rates, etc.)?

- **Population genetics**—A special branch of genetics. It does not concern itself with individuals except as packages of genetic material. Population genetics addresses questions such as the following:

  — What are the frequencies and distributions of specific genes within the population?

  — Are these frequencies stable or changing with time? How do they change with time? What causes these changes (mutation, natural selection, etc.)?

  — What are the mutation rates within the population? What are the rates by which gene frequencies change as a result of migration into or out of the population?

In this chapter, you will learn basic concepts of population biology as they apply to nonhuman and human populations. You will examine how human populations are growing and what effects this growth has on the environment.

WHAT IS A POPULATION?

A **population** is a group of the same species living in the same place at the same time. The place can be narrow or broad and is defined by the population biologist. For example, you could consider the earthworms that live in your garden as one population. Alternatively, the maple trees that make up a particular forest constitute another population. The human population on Earth is an example of yet another level of population. Because living organisms know no boundaries, the boundaries placed upon a population are usually arbitrary and defined by the scientist who studies that population.

Defining individuals within a population can be difficult. For animals, this is usually not a problem. For example, you can tell one rabbit from another because each is an individual that has its own anatomy, moves independently of others, and reproduces with other individuals of that population sexually. The rabbits are **units** of a population.

However, plants can present a problem in defining a population. Consider a forest in which the rabbits live. Is each tree in the forest an individual? Do the trunks represent individual trees? When trees become established, they may also reproduce asexually. Thus, some neighboring trees may be asexually reproduced offspring of one parent tree. They perform biological functions that are independent of the parent tree, yet they are genetically identical to the parent. Some of these trees are **modules**, that is, part of the whole. Therefore, to distinguish between these possibilities, biologists use the following terms:

- **Genet**—Individual plants produced from the zygote of sexual reproduction.
- **Ramet**—Individual plants derived from a parent by asexual reproduction.

A population consisting of ramets from the same parent is a **clone** and constitutes a single organism, regardless of how many "individuals" there appear to be.

HOW MANY INDIVIDUALS MAKE UP A POPULATION?

There are two basic ways to describe the number of individuals in a population: the actual size and the population density. The **actual size** of a population is the total number of individuals in the population. You can establish the total number by a direct count or **census** of the population. Often, you cannot count every member of a population, but you can estimate the actual size by counting representative samples and relating the proportion to the whole.

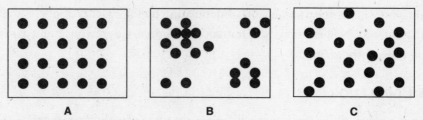

Figure 5.1 Possible Distributions of a Hypothetical Population:
(A) Uniform, (B) Clustered, (C) Random

Population density relates the number of individuals to a given unit of area. For example, the dandelion population in a field may be the number of dandelion plants per square meter. Human population density is often expressed as the number of people per square mile. Remember that some species live in environments with three dimensions (oceans, air, soil), so the population density might be the number of individuals per unit of volume, such as the number of phytoplankton/cm^3 or the number of fish/m^3.

While population density is a useful measure, bear in mind that populations may be distributed differently across any given area. There are three possible population distributions (see Figure 5.1):

1. **Uniform**—Individuals are equally spaced across a given area, such as crops in a field.

2. **Clustered** or **Grouped**—Individuals associate in small groups or clusters that are unequally spaced across a given area. Many animals are grouped in this manner.

3. **Random**—Individuals are not grouped in any particular pattern across an area. Windblown seeds may settle and establish themselves in an area in this manner.

Following are three basic methods used in measuring the size of a population (some specialized fields, such as fisheries management, use other methods):

1. Direct count or census

2. Quadrat method (sample plot method)

3. Capture-recapture

The **direct count method (census)** is just what it says: You count all of the individuals in the population. This method is done every 10 years by the U.S. government to establish the size of the population of the United States. Although every effort is made to count all citizens, it is not exact, because during the time it takes to conduct the census, families move within the country, people are born, and people die. So, at best, the U.S. Census is a reliable estimate of U.S. population size.

The **quadrat method** involves constructing a grid of sample squares, each of known area, within the larger area where the population lives. Individuals are counted in a fraction of those sample squares, and the total population size or density is then estimated. In some cases, the sample squares may actually be wooden squares that are distributed at random throughout the total area. In other cases, the sample squares may be roped off or imaginary squares (e.g., GPS coordinates). The quadrat method works best for counting organisms that do not move (e.g., plants, trees). Note that since the basic concept of this method is using a sampled area to extrapolate the entire population, the sample unit does not need to be square. Circles, droplets, or other shapes of known area or volume can also be used.

The **capture-recapture method** is used for estimating the size of mobile populations (e.g., animals). In this method, you count and mark a sample of individuals in the area on a given day and then return these individuals to the habitat. On the next day, you go back, count, and mark another sample of individuals (some of those individuals may have already been marked from the previous day and you note this number), and return them to the habitat. You continue this process over several days to estimate the population size. For example, suppose you want to know the population of Japanese beetles in a field. On Day One, you count and mark 100 beetles. On Day Two, you capture another 100 beetles and 20 of them are already marked. If you can capture a marked beetle as easily as an unmarked one, you can estimate the size of the population:

1. Day One: total captured = 100

2. Day Two: total captured = 100 (marked = 20); therefore, the fraction recaptured = 20/100, or 0.2

3. Estimate total size: number captured divided by the fraction recaptured

$$\text{Total size} = 100 \text{ captured}/0.2 = 500 \text{ beetles}$$

How Do Populations Grow?

To answer this question about population growth, let's first consider models of a nonhuman population in isolation, that is, a population wherein no immigration or emigration occurs. For example, suppose you place a package of yeast in a 1-liter flask of sugar water and make sample estimates of the population size over time. Ideally, if each yeast cell divides into two or more and successive generations continue, then the population will grow rapidly. The rate of change of the population or growth rate could be described by the following differential equation, where N is the population size, b is the instantaneous birth rate, d is the instantaneous death rate, and $\frac{dN}{dt}$ is the growth rate of the population:

$$\frac{dN}{dt} = (b-d)N \qquad \text{or} \qquad \frac{dN}{dt} = rN$$

In the second form on the right, r is the **intrinsic growth rate** ($b-d$), or reproductive rate. Basically, this equation says that the growth rate of the population will be proportional to the intrinsic growth rate and/or the population size. In other words, the faster the individuals reproduce, the greater the growth rate or the bigger the population, the faster the growth rate. If you plotted the population size versus time for such a population, it would look like Figure 5.2. This type of growth is called **exponential growth** and is considered indefinite growth or growth without limits.

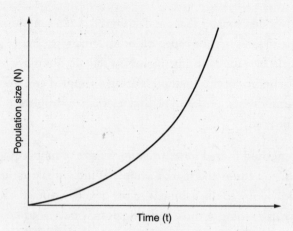

Figure 5.2 Exponential Growth Curve of a
Hypothetical, Ideal Population

However, real populations do not grow indefinitely. In the yeast example, the yeast cells might run out of food or space and stop growing. Ultimately, a population will grow until it reaches some type of limit, called the **carrying capacity** (*K*) of the environment. The equation for growth of this type is as follows:

$$\frac{dN}{dt} = rN\left(\frac{K-N}{K}\right) \quad \text{or} \quad \frac{dN}{dt} = rN\left(1 - \frac{K}{N}\right)$$

The unutilized portion of growth or resources is represented by $1 - \dfrac{K}{N}$. Basically, the equation states that a population will grow exponentially when *N* is low and most of the resources are unutilized. But as the unutilized portion of growth decreases (e.g., resources become low), the growth rate will slow. At the point when the population size (*N*) reaches the carrying capacity, the population will stop growing and remain constant. If you graphed this type of growth, it would look like Figure 5.3. This type of growth is called **logistical growth**, and the curve is often referred to as an "S-shaped curve."

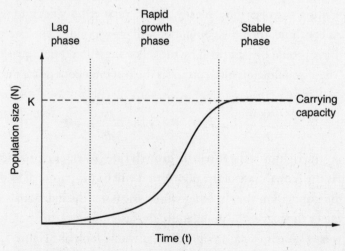

Figure 5.3 Logistical Growth Curve of a Hypothetical,
Real Population

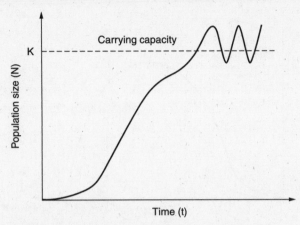

Figure 5.4 Logistical Growth Curve of a
Hypothetical, Real Population with Oscillations
around the Carrying Capacity

The logistical growth curve is characterized by the following three phases:

1. **Lag phase**—It takes a little time for the population to start growing.

2. **Rapid growth** or **Log phase**—The population is growing exponentially.

3. **Stable phase**—Rapid population growth slows as unutilized resources diminish and the population approaches carrying capacity.

As an investigator, you probably will not know the intrinsic growth rate (r) of the population that you are studying. You will probably know or have estimates of the population size (N) and, perhaps, the rate of change of the population ($\frac{dN}{dt}$). By plotting the population growth data and mathematically fitting the logistical growth equation to the data, you can obtain the value of r and even an estimate of the carrying capacity (K).

Oftentimes, populations do not stop growing completely when they reach the carrying capacity, but rather they oscillate around the carrying capacity. They will have spurts of growth and death around the carrying capacity, as shown in Figure 5.4.

Sometimes populations will grow to the carrying capacity and then, catastrophically, most of the population dies. Death may be the result of using all of the resources or disease. Such a growth curve is depicted in Figure 5.5.

So, what can set limits on population growth? The limitations can be either independent or dependent upon the density of the population. **Density-independent factors** are not related in any way to the size of the population at any given time. Examples of density-independent factors include the following:

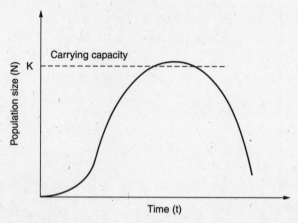

Figure 5.5 Logistical Growth Curve of a Hypothetical, Real Population That Catastrophically Dies after Reaching the Carrying Capacity

- Food supply

- Temperature changes in the environment (seasonal, climate)

- Weather (hurricanes, floods, drought)

- Habitat destruction (deforestation, human activities)

Let's look at some examples. Suppose a field of grass can support 10 rabbits, the carrying capacity. Now, a farmer fertilizes the field so that more grass can grow. Because there is more food available, the rabbits will reproduce and the population size will grow, perhaps double. Alternatively, if a wildfire destroys the field, then some of the rabbits will die of starvation and the population will decrease.

Density-dependent factors exert their effects as the population size increases. Examples of density-dependent factors include the following:

- Competition for resources (food, space)

- Predation

- Disease

- Changes in reproductive capacity—some species intrinsically slow down reproduction when their populations become overcrowded (rats, mice).

- Behavioral changes—some species resort to cannibalism and killing their young when they become overcrowded.

Let's look at some examples of density-dependent factors in the field of grass example above. Suppose the rabbits compete with field mice for the available grass. Each population (rabbits, mice) is limited by the available food. Now, suppose hawks flying above the field prey on the rabbits more than on the mice because the rabbits are bigger and easier to see; the rabbit population decreases. With the rabbit population reduced, more food becomes available to the mice and the mouse population increases (i.e., its carrying capacity is now greater).

Populations will differ in their rates of growth, but they fall into the following two major categories based on their intrinsic growth rates:

1. ***r*-strategists**—High birth rates, many offspring in a generation, and short life spans

2. ***K*-strategists**—Low birth rates, few offspring in a generation, and long life spans

Some *r*-strategists include insects, fish, bacteria, algae, and rodents. They have hordes of offspring that mature quickly (with little or no parental care), reproduce, and die quickly. *r*-strategists are opportunistic and can invade new niches and establish themselves rapidly. However, when environmental conditions change rapidly, their populations crash.

In contrast, *K*-strategists include sharks, birds, some reptiles, and most mammals. They have only a few offspring that take a long time to mature and require much parental care. They tend to be long-lived, live in stable environments, and maintain their population size near the carrying capacity.

As mentioned, species with various reproductive strategies have different life expectancies. Life expectancies can be represented in graphs called survivorship curves. (See Figure 5.6.)

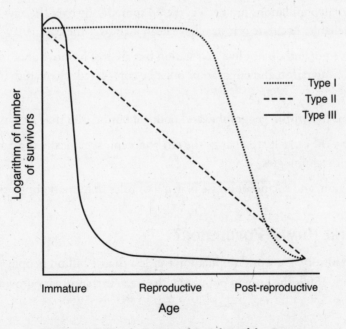

Figure 5.6 Basic Types of Survivorship Curves

Type I survivors are those with long life spans. The young survive, and death rates are high among the elderly. This type of curve is typical of mammals and some plants, such as redwood and other hardwood trees. **Type II survivors** have a fairly constant death rate at all ages. Many birds and flowering plants exhibit this type of survival curve. **Type III survivors** are characterized by high death rates among the young, such as marine invertebrates, insects, and fish.

To summarize models of population growth, real populations grow rapidly until they reach the carrying capacity of the environment. The carrying capacity is determined by interactions of both density-independent and density-dependent factors. Once the carrying capacity is reached, the population can become stable, oscillate around the carrying capacity, or crash catastrophically. The carrying capacity is not rigid, but dynamic; it can change as environmental conditions change. Finally, in any given environment, the carrying capacity for one population is not necessarily the same as the carrying capacity for another.

THE DYNAMICS OF HUMAN POPULATION

Does the human species follow the same biological pattern with respect to populations as other organisms? Is there a limit to human growth? If so, what is the carrying capacity of Earth for humans and what affects it? When it comes to humans, these are complex questions that are difficult to answer for the following reasons:

- Data and principles of population biology have been derived from laboratory experiments and field observations of nonhuman species. Their relevance to humans is hotly debated.

- Data on human populations has been gathered sporadically over the ages and much early data are unreliable. Accurate censuses have been kept over the past 100 years or so.

- Humans have not and do not live in isolation but are free to move about the planet. Therefore, immigration and emigration must be considered when studying human populations.

- Most humans live within geographic and political boundaries that are largely artificial.

- Humans have the capacity to change the environment, specifically the carrying capacity, through their technologies.

- Politics, religion, and socioeconomics play major roles in human population decisions.

How Large Is the Human Population?

For most of human history, the world population was less than 1 billion people (Figure 5.7). However, since about 1800, the human population has seen extraordinary growth. Current

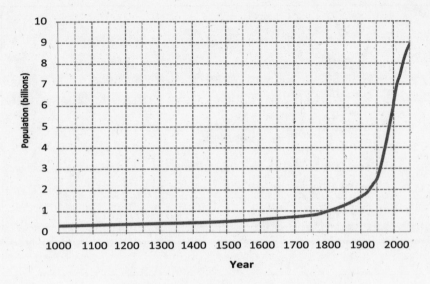

Figure 5.7 World Population, 1000–2050

estimates place the human population at over 7 billion people, and projections of the current growth rate will place it at over 9 billion by 2050.

The human population is not evenly distributed across the globe but is clustered throughout the continents in many countries. The continents of Asia, Africa, and Latin America have the highest populations. The five most populous countries from highest to lowest are China, India, the United States, Indonesia, and Brazil. (See Figure 5.8.) Even within countries, people are clustered. For example, in the United States, the major population centers are along the Northeast coast, the Pacific coast, and the central Midwest along the Great Lakes.

How Do You Study the Growth of the Human Population?

The data for human population studies must be obtained by reliable direct counts or census data. Thus, demography is important. For any given region (most likely countries), you must have reliable estimates of population size, number of births, number of deaths, immigration numbers, emigration numbers, number of males, number of females, and ages of the members of the population.

While it seems reasonable that humans should follow a logistical growth curve like other nonhuman species, we do not typically apply those models to human population growth. Instead, we use models that consider the initial population size, the factors that add to the population (births, immigrations), and the factors that take away from the population (deaths, emigrations). So an equation to describe the population is as follows:

$$N_t = N_o + (B + I) - (D + E)$$

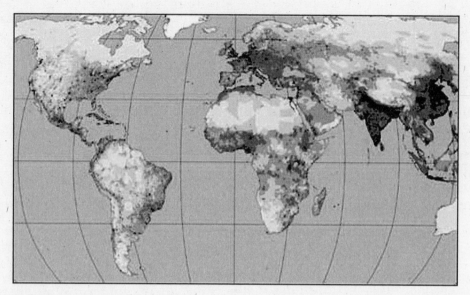

Figure 5.8 Distribution of the Human Population across the World. The gray areas represent dense human populations. The darker the gray, the denser the population.

Source: Courtesy of NASA, available at http://science.hq.nasa.gov/research/daac/sedac.html.

Where N_t is the population size at any time, N_o is the initial population size, B is the number of births, I is the number of immigrants, D is the number of deaths, and E is the number of emigrations.

We can write an equation to calculate the change in population with time as (ΔN_t) as follows:

$$\Delta N_t = (B + I) - (D + E)$$

These equations are useful for characterizing changes in human populations by using census data, as we often do not know what the carrying capacity of the global environment for humans is.

So, if we do not typically apply logistical growth models, then how else can we characterize the parameters of human growth? Is there a human equivalent to intrinsic growth rate (r)? When we talk about human growth, we refer to four parameters:

1. Crude birth rate

2. Crude death rate

3. Annual rate of population change

4. Doubling time

Let's look at each of these parameters.

The **crude birth rate** is the number of live births per thousand persons in a population in a given year. According to the U.S. Census Bureau's population clock, the number of world births as of March 2006 was 130.9 million. If we divide that number by the world population of 6.5 billion (6,500,000 thousand), we get a crude birth rate for the world of 20.1 births/1,000.

The **crude death rate** is the number of deaths per thousand persons in a population in a given year. According to the U.S. Census Bureau's population clock, the number of world deaths as of March 2006 was 56.6 million. If we divide that number by the world population of 6.5 billion (6,500,000 thousand), we get a crude death rate of 8.7 deaths/1,000.

The **annual rate of population change (ARPC)** is expressed as a percentage according to the following equation:

$$\text{ARPC (\%)} = \frac{\text{birth rate} - \text{death rate}}{1,000} \times 100$$

This equation simplifies to the following:

$$\text{ARPC (\%)} = \frac{\text{birth rate} - \text{death rate}}{10}$$

So, as of March 2006, the ARPC = $(20.1 - 8.7) \div 10 = 1.14\%$.

To find the actual number of the increase in population over a year, multiply the ARPC by the population size. For 2006, you will get 74.1 million. The ARPC is analogous to the intrinsic growth rate (r) calculated by the logistical growth curve models for nonhuman populations.

The change in ARPC over time is shown in Figure 5.9. The ARPC has been decreasing since 1965. Although the ARPC has been decreasing, this does not mean that the human population is not growing. It only means that the rate of growth is slowing. For the population to remain constant, birth rates must equal death rates (i.e., **zero population growth**) and the ARPC must equal zero. For the population to decrease (i.e., **negative population growth**), the death rates must exceed the birth rates, and, therefore, the ARPC must be a negative number.

Another way to express the rate of population growth is by calculating the doubling time. Doubling time is the number of years that it will take a population to double if the current growth rates remain constant. Doubling time is calculated by the following rule of 70:

Doubling time = 70 years/annual rate of population change (%)

As of 2006, the doubling time of the world human population would be as follows:

Doubling time = 70 years/1.14 = 61.4 years

Note that as the ARPC increases, the doubling time shortens. Also note that the calculation of doubling time assumes a constant growth rate, which is not necessarily always true. However,

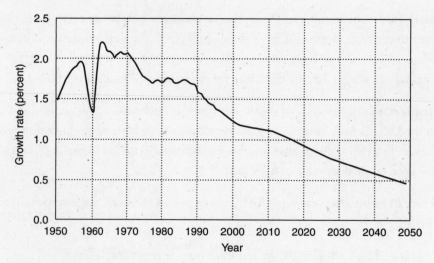

Figure 5.9 The Annual Rate of Population Change (ARPC), 1950–2050
Source: Courtesy U.S. Census Bureau, available at http://census.gov/ipc/www/img/worldgr.gif.

doubling time is a useful rule of thumb calculation to estimate a population's growth rate and for projecting how a population will impact its resources.

WHAT FACTORS AFFECT GROWTH RATES OF HUMAN POPULATIONS?

Before we consider the factors that affect the growth of human populations, we must note several things. First, because we are dealing with humans, these factors are not strictly biological as they are with nonhuman species. Many factors that influence human population growth also deal with politics, economics, religion, and social mores and customs. Second, the human population is not one cohesive unit, but rather it is broken up into countries. Third, countries are also categorized into the following geopolitical/economic groups:

- **Industrialized countries** (originally classified as the First World)—United States, Canada, Great Britain, Germany, France, Japan, etc.

- **Moderately developed countries (MDCs;** originally the Second World)—China, Russia, former Soviet bloc countries in Eastern Europe, etc.

- **Less developed countries (LDCs;** originally the Third World)—India, African nations, Latin American nations, etc.

These labels carry stereotypes and attitudes about socioeconomic development, but they are still widely used.

From our model for human growth [i.e., $\Delta N_t = (B + I) - (D + E)$], you can see that the growth rate will be affected by factors that add or subtract from the population. These factors include the following:

- Birth rates
- Death rates
- Migration (immigration and emigration)

BIRTH RATES

Crude birth rates are expressed per thousand people in the population. However, the population includes women, men, children, and the elderly. Because only women between 15 and 49 years generally can bear children, the population growth might better be expressed by measures of **fertility**, the actual bearing of offspring. Fertility rates are the number of children born per number of women (usually 1,000). Following are four ways to express fertility:

1. **General fertility rate**—Number of live births/1,000 women/year
2. **Age-specific fertility rate**—Number of live births/1,000 women of specific age/year
3. **Total fertility rate**—Number of children a woman will bear during her reproductive life
4. **Replacement-level fertility**—Number of children a woman must bear to replace a set of parents (2.1 is generally considered replacement level fertility)

Fertility rates tend to be higher in LDCs and MDCs than in industrialized nations. For example, replacement fertility in MDCs and LDCs tends to be around 2.5, as these regions have higher rates of infant mortality.

Factors that influence birth rates and fertility rates include the following:

- **Infant mortality**—Infant mortality is defined as the number of babies out of every 1,000 born each year that die before their first birthday. In areas where infant mortality is high (LDCs), birth and fertility rates are high.

- **Marriage age**—In countries where women tend to marry at a young age (below 25 years old), birth and fertility rates tend to be higher.

- **Education**—Fertility and birth rates tend to decrease as the level of education increases.

- **Affluence**—Fertility and birth rates tend to be lower in affluent countries.

- **Child labor**—In countries where children are a major part of the labor force, birth and fertility rates tend to be higher. However, countries with high rates of child labor also tend to be the least affluent and least educated.

- **Opportunities for women**—When women have access to education and employment, birth and fertility rates tend to be low.

- **Availability of birth control**—Widespread availability and use of birth control methods reduce birth and fertility rates.

- **Religious/cultural beliefs**—In countries where large families are deemed desirable, birth and fertility rates are high.

DEATH RATES

The increase in human population growth over the past 100 years or so has not been caused by an increase in birth rates as much as a decrease in death rates. Crude death rates have decreased mainly for the following reasons:

- **Nutrition**—Better nutrition and increased food production have reduced deaths due to starvation.

- **Sanitation, water, and hygiene**—Improvements in these areas have reduced infant mortality and lengthened life expectancies.

- **Medicine and public health**—Improvements in medical technology and health care delivery (antibiotics, immunizations, insecticides) have reduced deaths caused by many diseases (smallpox, measles, influenza, etc.).

A useful indicator of reduced death rates is **life expectancy**, that is, the average number of years a newborn can be expected to live. Life expectancy varies with socioeconomic conditions and therefore tends to be lower in MDCs and LDCs compared to industrialized nations. Currently, the average life expectancy in industrialized nations is 85 years, up from 75 years in 1965.

MIGRATION

Migration into or out of a country can greatly influence the population size. For example, in the United States, a large part of the population growth increase reflects immigration rather than changes in birth and death rates.

WHAT IS THE CARRYING CAPACITY FOR HUMANS?

In nonhuman populations, the environment sets limits on the growth of any one species. This limit is the carrying capacity and is determined by density-dependent and density-independent factors. Although scientists agree that there is a limit to human population, no one equation to calculate the carrying capacity can be agreed upon. Several other factors come into play:

- Humans live within geopolitical borders, not necessarily environmental ones. Natural resources, such as rivers, forests, and lakes, can cross national boundaries. So, by utilizing common resources, one country can affect another country's population (e.g., damming rivers, burning forests, etc.).

- Humans can raise the carrying capacity of the environment through advances in technology, especially agriculture.

- Carrying capacity in nonhuman populations is phrased in terms of survival. However, for humans, it is better thought of as standard of living. Countries with a high standard of living utilize more resources than those in countries with lower standards of living. Therefore, those countries with high standards of living tend to have lower populations.

- When nonhuman populations exceed the carrying capacity, members of the population die off. If any one country exceeds its resources, other countries can offer aid to that population and maintain it.

So, from a strictly biological point of view, it is difficult and complex to try to estimate the human carrying capacity.

How Are Human Populations Structured?

As mentioned at the beginning of this chapter, demography is the field that describes the statistics of populations. Demographic data are extremely important in human population biology and making policy decisions about human populations. Demographic data come from censuses, and data are broken down into **vital statistics** (age groups, sex, economic status, geographic regions, etc.).

A major way to describe a country's population is by an **age-structure diagram**, as shown in Figure 5.10. An age-structure diagram is also known as a **population profile, age-structure pyramid**, or **population pyramid**.

An age-structure diagram shows the makeup of a population by age and sex. On the x-axis is the percent of the total population broken down by males (on the left) and females (on the right). The y-axis is broken down into specific age categories. For example, in Figure 5.10, both males and females under age 5 make up about 3.5 percent of the population. You can tell the following things from an age-structure diagram:

- **Sex ratio**—This is the ratio of males to females and can be determined by symmetry about the y-axis. For the most part, in the United States, the sex ratio of males to females is about even (the diagram is symmetrical about the middle axis). However, at about age 60 and above, females begin to outnumber males, reflecting the longer life expectancy of females compared to males in the United States.

- **Population booms or growth spurts**—If you look at the profile of the United States, you will see a bulge between the ages of 35 and 55. This represents the post–World War II baby boom, when there was a huge increase in the population. If you plot population pyramids over several decades, this bulge will move up the pyramid as the baby boomers mature.

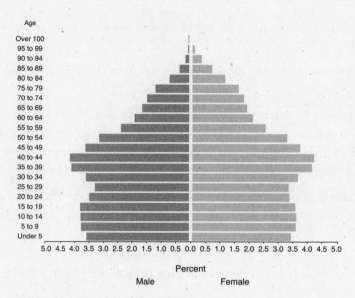

Figure 5.10 Age-Structure Diagram of the U.S. Resident
Population as of July 1, 2000

Source: Courtesy of U.S. Census Bureau.

- **Dependency loads**—The dependency load is the portion of the population that is under age 15 and over age 65. These people are generally considered either too young or too old to work and contribute to the nation's gross national product, but they do consume a nation's resources.

Age-structure diagrams exhibit the following patterns of growth rate (see Figure 5.11):

- **Rapid growth** is characterized by a pyramid shape with a large base and narrow peak. These populations have high proportions of children (high birth rates) but lower proportions of adults, especially the elderly (high death rates). Infant mortality is high and life expectancy is low (e.g., LDCs such as African nations and India).

- **Slow growth** is characterized by an almost triangular-shaped pyramid. Birth rates are higher than death rates, but there are more adults than children. Life expectancies are greater than in rapid growth populations (e.g., industrialized nations and MDCs such as the United States and Canada).

- **Zero growth** is characterized by an almost rectangular-shaped pyramid. There is a large proportion of adults compared to children. Life expectancies are greater than in rapid growth populations (industrialized nations and MDCs such as Italy and Austria).

- **Negative growth** is characterized by a shallow trapezoid with a narrow base. The proportion of people at nearly all ages is about the same, but the birth rates are much less and the population is declining (industrialized nations such as Germany).

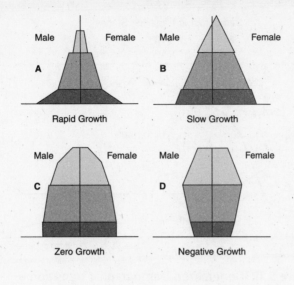

Figure 5.11 Age-Structure Diagrams for Countries
with Various Population Growth Rates

When demographers analyzed the development and industrialization of European countries in
the 1800s, they noticed a pattern of population change and developed a hypothesis known as
demographic transition. Basically, demographic transition means that, as countries become
industrialized, the birth rates and death rates decline. The transition takes place in the following
four stages (see Figure 5.12):

1. **Pre-Industrial**—High birth rates, high death rates, and low population growth rates are
   typical.

2. **Transitional**—Death rates drop due to improved health care and increased food
   production. Birth rates remain high, so the population grows rapidly.

3. **Industrial**—Birth rates drop and approach the death rates, and the population growth
   slows. (Most MDCs are in this phase.)

4. **Post-Industrial**—Birth rates decline to at or below the death rates (negative population
   growth), and the population size decreases.

In some cases, rapid growth rates and deterioration of the environment reduce the standard of living
and impoverish the population. Poverty brings a sense of helplessness to the citizens, and they rely
on large families (more children) to help their economic situation. These conditions prevent the
country from progressing out of stage 2 and have been referred to as a **demographic trap**.

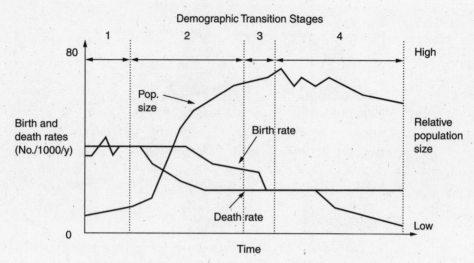

Figure 5.12 Generalized Demographic Transition

REGULATING POPULATION SIZE

If you look at the equation for the rate of change of human population growth [$\Delta N_t = (B + I) - (D + E)$], you will see that only the following few factors are involved in regulating population size:

- Birth rate
- Migration (immigration, emigration)
- Death rate

If you want to increase the population size, you can increase birth rates, decrease death rates (e.g., through improved health care), and allow more immigration. (Most countries do not control emigration—that is left up to another country's immigration laws.) If you want to decrease the population size, you can reduce birth rates and immigration (increasing death rates is not ethical).

Many countries have instituted **population policies** that mostly regulate births or immigration. The idea is to match the population size with the country's resources or its standard of living. The population policies that regulate births or family sizes fall into one of two categories:

1. Pronatalist policies
2. Antinatalist policies

Pronatalist policies encourage births and large family sizes so that the population can grow. In the 1930s, France adopted a pronatalist policy that paid families government subsidies for raising children and paid a maternal salary to women who had additional children beyond two (the replacement level). Some Eastern European countries (former East Germany, Romania) had new home loan policies to encourage young parents to have children and policies to forgive some portion of the debt with each new child born. Romania banned abortion so that no woman under 45 years old with fewer than five children could get an abortion. (This law was repealed in 1989.)

Antinatalist policies discourage births and encourage small family sizes so that the population will become stable or even decline. China and Singapore instituted such policies as raising the legal age of marriage to shorten a woman's period of fertility. In Singapore through the 1980s, for instance, low-cost housing or housing subsidies were reduced when women had more than two children. The decisions on which policies to enact depended upon the way population growth was perceived relative to the standard of living and available resources (food, housing, sanitation, water supplies, health care, etc.).

Some people believe that population growth should not be regulated. To them, decisions regarding number of children and family size are basic rights that are governed by an individual's religious beliefs, personal privacy, and freedoms. In contrast, others believe that population growth should be regulated. Regarding population-growth regulation, there are two basic philosophies:

1. Oppose any population growth
2. Oppose rapid population growth

Those who oppose any population growth believe that any population growth will strain the world's resources and lead to environmental deterioration. This deterioration will ultimately lead to a reduced standard of living, lower the carrying capacity, and increase violent conflicts as countries compete for limited resources. This group advocates public policies, educational programs, and laws to limit birth rates to replacement levels or below. In contrast, some people oppose rapid population growth, especially in LDCs, because it is perceived as having negative impacts on economic development and standard of living. This group advocates economic and social development of LDCs in the hopes that as living standards and education improve, parents will not see a need for large families.

So what can be done to limit population growth? Following are three major avenues:

1. Family planning
2. Social and economic incentives
3. Policies and laws

Family planning includes a wide variety of methods that parents can use to control the number of children that they have as well as the timing between children. The goals of family planning are not so much to limit births as to enable families to take care of the children they have and to keep them healthy (if you can keep a smaller family healthy with low infant mortality, you don't need a large family). Family planning programs focus on education in the following areas:

- **Human sexuality**—How are children conceived? When is the appropriate age to have children? What are the risks and responsibilities of having children?

- **Childcare**—Who provides prenatal and postnatal childcare? What is the role of good nutrition for the mother before conception and during pregnancy? What is the role of breastfeeding in infant childcare?

- **Contraception**—What contraceptive methods are available (condoms, birth control pills, sterilization, etc.)? Which methods, if any, are socially and religiously acceptable in any given country?

In some countries, family planning clinics are essential contributors to health care where women can get medical services (prenatal check-ups, postnatal check-ups, distribution of contraceptives, sterilization procedures, abortions) when necessary.

Social and economic incentives usually include subsidies or aid from governments. Aid could be in the form of actual payments, tax relief, debt relief, low-cost housing, food, or medical care. Generally, families with fewer children receive more benefits than those with a greater number of children (i.e., below replacement level versus greater than replacement level). In some cases, penalties can be assessed for having more children than replacement level (fines or withdrawal of benefits).

Policies and laws include antinatalist policies as discussed above. The extent of such laws and policies depends upon the country. Let's look at some examples.

UNITED STATES—A SLOW GROWTH RATE COUNTRY

The United States has no official population policy because decisions regarding family size are considered the rights of individual families. Laws regulating family size and fertility would be construed as an intrusion by the government on individual liberties. However, the following laws and social mores tend to favor population growth by promoting family sizes:

- **Tax laws**—U.S. income tax laws have deductions for families and child tax credits.

- **Abortion laws**—There are legal limits to abortion and limits regarding federal funding for abortion procedures.

- **Welfare laws**—child benefits

- **Sex education**—Sex education in public schools is regulated at the local and state level and is often influenced by social and religious beliefs. Some areas encourage comprehensive education, while other areas have more limits on their programs.

The major factor contributing to the U.S. population growth is not increased birth rate, but rather increased rates of immigration, both legal and illegal.

INDIA—A RAPIDLY GROWING POPULATION

In 1952, India had a rapidly growing population of 400 million and a large percentage of impoverished citizens. The democratic government instituted population policies aimed at reduced growth. These policies included increased family planning, education, age of marriage, and employment. However, much of India's population is in rural areas, and social beliefs there encourage large families. Despite these control efforts, India's population exceeded 900 million by the 1990s. Factors that have contributed to the population policy's failure in this period include poor planning, inefficiency of implementation, poverty, low social status of women, and lack of financial support. In the 21st century, India's economy is improving, which may lead to a reduction in population growth, but this has yet to be determined.

CHINA—STEMMING A RAPIDLY GROWING POPULATION

China has over 1 billion people, which is about one-sixth the world's population. But China has only about 7 percent of the world's farmable land. Clearly, a mismatch exists between population size and resources (land, water, food production capacities), which is also reflected in the standard of living. Up until the first half of the 20th century, Chinese people generally had large families for economic reasons. Recognizing these facts, the Chinese government instituted the following policies in the 1960s and 1970s:

- **Extensive health care**—This program included China's rural poor.

- **Education and employment**—Opportunities were extended to women.

- **Social Security**—Old age pensions and food cooperatives meant that the elderly did not necessarily have to rely on their families for a living.

- **Marriage age**—Laws increased the age of marriage to the early 20s.

- **Family planning policy** (unofficially known as the One-Child Rule)—While the government began providing contraception to the population in the 1960s, starting in 1979, most Chinese couples were limited to one child. Fines or other penalties, including sterilization or abortion, were levied on couples having additional children. There are various exemptions, and the policy was accompanied by education promoting delayed childbearing and greater access to contraception.

These policies are aimed at improving the standard of living of the Chinese people and, in so doing, reducing the rate of population growth. One important factor is that the Chinese government is a totalitarian government that has the authority to implement these policies with or without the public consent. The Chinese people are being asked to temporarily put aside individual desires for the good of the state.

JAPAN—A SUCCESS IN REDUCING POPULATION GROWTH

After World War II, Japan successfully reduced its rapid population growth by implementing family planning policies (liberal abortion law, access to contraceptives), improving health care, and improving pensions. The success of the Japanese economy also contributed to the success of these policies. However, as Japan approaches zero growth, one problem it is experiencing is the aging and overall decline of the workforce. To solve this problem, Japan has automated many industries, thereby reducing the number of workers needed. Another problem in Japan is that an aging population will place a great economic demand on the government's health care and pension systems while reducing its tax base. This problem has yet to be resolved.

In summary, to manage human population growth the following things must be done:

1. Recognize the "carrying capacity" of the environment. What population can the country's resources support to provide a decent standard of living for all?

2. Make decisions regarding the desired rate of population growth—match the available resources with the rate of population growth.

3. Implement family planning, education, and socioeconomic programs to achieve the goals. Most of these programs are geared to increase the standard of living, employment, and security, while reducing the birth rates (directly or indirectly).

4. Conserve the environmental resources as much as possible.

THE IMPACT OF POPULATION GROWTH

In both nonhumans and humans, increasing population sizes impact the environmental resources (food, space, water, etc.). While nature regulates population growth in nonhuman species, governments must regulate human population growths. But what are the recognizable impacts of human population growth on the environment and on the population itself? Impacts include the following:

- Hunger
- Poverty
- Disease
- Depletion of resources (water, energy, minerals)
- Living space (urbanization)
- Habitat destruction

Let's look at these impacts.

HUNGER AND POVERTY

Every human needs a healthy diet to provide energy for growth, development, work, exercise, and other biological processes. The diet should be diverse in carbohydrates, fats, and proteins, and should be about 2,400 calories per day. While domesticated animals provide proteins, the bulk of the world's diet comes from grains (wheat, rice, corn, barley, potatoes, etc.).

Food production requires land for agriculture (11 percent of Earth's land surface is usable for crops). In addition to land, some food production (fish) comes from freshwater bodies or the ocean. Agricultural land is limited and unevenly distributed among countries. Despite this, advances in agricultural technology have greatly increased the amount of food that can be produced per acre. So, many economists believe that, with proper management and investment, the available agricultural land is capable of meeting the food needs of the human population. Yet, even with this optimism, many of the world's inhabitants die of hunger, starvation, and malnutrition (about 40 million people per year).

But what are hunger, starvation, and malnutrition? Let's look at some definitions:

- **Hunger**—an aching desire for food that goes unabated for days, weeks, and even longer
- **Starvation**—suffering or death from being deprived of nourishment
- **Famine**—widespread starvation
- **Undernutrition**—chronic consumption of too few calories per day
- **Malnutrition**—too little consumption of specific nutrients, which often leads to deformations
- **Malabsorptive hunger**—as a result of undernutrition and/or malnutrition, the body loses the ability to absorb nutrients from consumed food.

Hunger and malnutrition are common in many LDCs. Generally, people in these countries consume about 10 percent of the required daily calories or less, while those in industrialized nations consume more than the required calories daily. Even in industrialized nations, hunger exists in various areas.

Hunger is related to poverty. In impoverished countries, individuals may spend up to 70 percent of their daily income to purchase food. Poverty is part of a vicious five-part cycle:

1. Poverty leads to undernutrition and malnutrition.

2. Malnutrition leads to decreased energy and increased infant mortality.

3. Decreased energy leads to decreased resistance to diseases.

4. Decreased resistance to diseases reduces the life expectancy and the ability to learn and work.

5. The reduced ability to work contributes to more poverty. The cycle repeats itself.

So, if we have the technology to produce enough food for the world's population, why do people go hungry? Hunger exists for the following reasons:

- A significant fraction of food production goes to feed domestic animals.

- Some areas of the world are more fertile than others.

- Many agricultural lands are devoted to cash crops (e.g., coffee, tobacco, cotton). In some cases, countries continue to produce cash crops even when significant numbers of their citizens are dying of famine.

- Land ownership is unevenly distributed between the wealthy and the poor. Many poor people work the lands of wealthy landlords, rent land at high prices, and/or make a living on marginally fertile lands.

- Poor countries may not have or be able to afford new agricultural technology.

DISEASE

As population size increases and exceeds its resources, the citizens become susceptible to diseases for the following reasons:

- Many people become impoverished. Poverty leads to hunger that, in turn, reduces resistance to diseases.

- High populations place increased demands for health care that the nation's economy may or may not be able to support.

- Increased population sizes also place demands on the availability of clean water and sanitation. Poor sanitation leads to diseases such as cholera, dysentery, and hepatitis.

- Diseases can spread more rapidly among dense populations than sparse ones (epidemics, pandemics).

DEPLETION OF RESOURCES

Besides food production, countries also have water, energy, and mineral resources. Increases in human populations, both locally and globally, place increased demands on resources.

Water   Freshwater must come from lakes, rivers, and streams. Obviously, increased local populations will increase the demands for freshwater. Often, different countries share these reservoirs. The activity of one country can affect that of another. Let's look at the example of a river that is shared by two countries, one upstream and the other downstream. The two countries start with equal population growths. Suppose the upstream country has a rapid burst of population growth and increases demand upon the water supply from the river. By taking more water, the upstream country reduces the water available to the downstream one. The downstream country's population and agriculture suffer from a lack of water (i.e., a new limit has been placed upon them). For example, Southern California has a large population and gets a major portion of its water (drinking, agriculture) from the Colorado River. Over time, the increased demand has reduced the flow of the river into Mexico, where it empties into the Gulf of California, thereby making less water available to the inhabitants of the Gulf of California region of Mexico.

In addition to consumption and agriculture, water is used for sanitation. Increased population growth places demands on sanitation and increases the discharge of human wastes into rivers and streams. This discharge may also reduce potable water supplies downstream.

Energy   Human populations require energy for electricity, heating, and transportation. The energy sources important for these tasks are nonrenewable sources such as wood and fossil fuels (coal, oil, natural gas). As a nation's population increases, so do the demands for these energy resources. By far, industrialized nations are the greatest consumers of these resources. Yet in recent years, as more countries enter into industrialization (e.g., China, India) and as populations grow in these countries, the demand for fossil fuels has increased.

The sources of fossil fuels vary. The United States has large coal reserves in the Appalachian region, and oil/natural gas in Alaska, Texas, Oklahoma, Louisiana, and offshore in the Gulf of Mexico. Worldwide, oil and natural gas supplies are abundant in Latin America (Mexico, Venezuela), Africa (Libya), and the Middle East (Saudi Arabia, Iraq, Kuwait, Iran).

Growing nations also exert increasing demands on the world supplies of oil and gas. As oil and gas are nonrenewable energy sources, the supply will eventually dwindle; there is much debate as to when this will actually occur. Furthermore, increased demands have driven the price up and redistributed wealth to those areas rich in oil (e.g., Latin America, the Middle East). This demand and supply has fueled much political unrest in these areas of the world including dictatorships and wars.

Minerals   Minerals (iron ore, aluminum, coal, nitrogen, phosphorus, platinum, titanium, gold, etc.) are used in many industrial processes (steel production, fertilizer production), manufacturing, and construction. Because mineral deposits are linked to global geological processes (plate tectonics), they are not evenly distributed across the globe. Some countries are mineral-rich, while others are mineral-poor. When population sizes increase, there is increased demand for manufactured goods, thereby increasing the demands on mineral resources. Existing resources must be used and new ones found. When mineral deposits are found, they must be extracted. Mineral extraction can have detrimental effects on the environment (habitat destruction, pollution, land use, etc.).

LIVING SPACE

As mentioned earlier, the world's population is not evenly distributed across the globe. (See Figure 5.8.) In most countries, the urban areas (cities, towns) are densely populated, while rural areas are less populated. As populations grow, people will move into urban areas in search of jobs and better living conditions. As a result, urban areas are growing rapidly as they attract people in search of jobs (61 percent of the world's population is projected to live in urban areas by 2025). As urban areas become more densely populated, the inhabitants place more demands for living space and basic services (food, water, sanitation, etc.). If these needs are not met, then the urban areas will decay into poor slums and will become impoverished and disease-ridden. As cities grow, they take up more land around them, which results in habitat destruction. In many industrialized countries, many cities merge to form huge megalopolises. For example, the United States has two such megalopolises in the east, the Boston to Washington, DC, corridor and the Pittsburgh to Chicago corridor. As cities grow, land use and planning become important factors to handle the increased needs for services.

HABITAT DESTRUCTION

In any country, as the human population grows, it places ever-increasing demands on the environment, such as the following:

- **Food**—Land must be cleared for agriculture or grazing, which destroys forest and grassland habitats that are used by other organisms. Overfishing reduces food supplies in offshore fisheries.

- **Water**—Rivers, lakes, and streams can be drained to provide drinkable water for growing populations.

- **Living space**—To provide space for expanding cities and roads to connect cities, areas of forests and grasses must be cleared and wetlands filled in. This clearing also destroys habitats for many nonhuman species. Increased construction usually leads to increased demand for wood products from forests, leading to deforestation.

- **Mineral extraction**—In mineral-rich areas, mining techniques (e.g., strip mining) often destroy habitats.

While these demands are directly related to consumption, humans also produce wastes. One of these is human sewage that gets dumped into rivers and streams, thereby polluting and degrading multiple habitats downstream. Another is human refuse that either gets burned or buried. Toxic chemicals from the refuse can make their way into the air or groundwater and degrade ecosystems near and far. Finally, air and water pollution from human industries and manufacturing can degrade ecosystems, local and far away (e.g., acid rain). So, increasing the human population places higher consumption demands as well as ecological stresses on habitats and ecosystems worldwide.

REVIEW QUESTIONS

1. An age-structure diagram has the shape of a pyramid with a wide base. What type of population growth does it indicate?

   (A) Zero growth
   (B) Negative growth
   (C) Slow growth
   (D) Rapid growth
   (E) Unknown growth

2. Rapid growth in nonhuman populations occurs in which phase?

   (A) Lag phase
   (B) Stable phase
   (C) Carrying capacity
   (D) Oscillating phase
   (E) Log phase

3. The U.S. Census Bureau has the following information on the populations of five countries for last year. The data are shown below:

	Birth Rate (No./1,000/yr.)	Death Rate (No./1,000/yr.)	Immigration Rate (No./1,000/yr.)	Emigration Rate (No./1,000/yr.)
Country A	20	20	10	5
Country B	10	5	20	10
Country C	30	20	5	15
Country D	65	50	10	10
Country E	10	5	10	10

Which country has zero population growth for last year?

   (A) Country A
   (B) Country B
   (C) Country C
   (D) Country D
   (E) Country E

4. All of the following factors are density-independent factors EXCEPT which one?

   (A) Competition
   (B) Climate
   (C) The pH level
   (D) Water supply
   (E) Food supply

5. Which of the following organisms would be an example of an *r*-strategist?

   (A) Whale
   (B) Cockroach
   (C) Eagle
   (D) Alligator
   (E) Dog

6. Which provision might BEST be used as part of an antinatalist policy in a country whose religious beliefs are against contraception?

   (A) Legally mandated sterilization after two children
   (B) Making condoms widely available
   (C) Tax breaks for large families
   (D) Raising the legal age of marriage
   (E) Free birth control pills

7. Your country has recently won a war, and the soldiers have returned home to their wives and families after many long years. What might the age-structure diagram look like if plotted over the next several decades?

   (A) A pinch at the bottom that will rise over the next several decades because of a boom in population
   (B) There will be no change in the shape of the age-structure diagram
   (C) A bulge at the bottom that will rise over the next several decades because of a bust in the population
   (D) A pinch at the bottom that will rise over the next several decades because of a bust in the population
   (E) A bulge at the bottom that will rise over the next several decades because of a boom in the population

8. You seed two flasks containing nutrient media with the same amount of the algae, Chlorella. Flask B contains twice the concentration of phosphate, a vital nutrient, than Flask A. Both flasks are incubated under the same conditions. Which of the following statements will be TRUE?

   (A) Flask A will grow faster than Flask B because the intrinsic growth rate is greater.
   (B) Both flasks will grow at the same rate and reach the same carrying capacity.
   (C) While the intrinsic growth rates are the same, Flask B will grow at a faster rate and reach a greater carrying capacity than Flask A.
   (D) While the intrinsic growth rates are the same, Flask A will grow at a faster rate and reach a greater carrying capacity than Flask B.
   (E) Flask B will grow faster than Flask A because the intrinsic growth rate is greater.

9. City officials face an increasing population size. They decide to fill in wetlands upstream along a river that runs through the city. They build homes on the filled-in land. As a result, they must periodically deal with floods along the river that brings water into the city. This is an example of which of the following?

   (A) Pronatalist policy
   (B) Antinatalist policy
   (C) Habitat destruction
   (D) Pollution
   (E) Ecological stress

10. In the transitional phase from Pre-
    Industrial to Industrial economies, birth
    rates _____ while death rates
    _____.

    (A) do not change; drop
    (B) increase; remain the same
    (C) decrease; decrease
    (D) decrease; remain the same
    (E) increase; increase

FREE-RESPONSE QUESTION

You are a demographer for the United Nations and have been assigned to analyze the country of Florin. Florin is a small, mountainous country in Eastern Europe. About 20–30 percent of the land is available for farming. The major crops are wheat, barley, and grapes. Much of the barley and grapes is used to make beers and wines for export. The majority of the citizens work in the vineyards and fields of a few wealthy landowners and subsistence farm on small plots of land where they live. The majority of the citizens are below the poverty level. The country's main religion is Catholicism. They believe in large, extended families and children are heavily involved in the labor force. The citizens have limited, minimal health care available to them. Although officially a democracy, the government of Florin is more totalitarian with a weak president and the ruling body made up primarily of the wealthy landowners. In an effort to improve the standard of living for the Florinese people, the government has asked for help from the United Nations. Florin has a small but growing population of about 100 million people. The following demographic data are available for Florin:

Age (yrs.)	Percent of Total Population	
	Male	Female
0–4	9	9
5–9	8	7
10–14	7	6
15–19	5	5
20–24	4	4
25–29	3.5	3.7
30–34	3	3.1
35–39	2	2.4
40–44	2.2	1.9
45–49	1.8	1.8
50–54	1	1
55–59	0.9	0.8
60–64	0.7	0.7
65–69	0.5	0.6
70–74	0.2	0.3
75+	0.1	0.2

(A) **Discuss** the type of population growth that Florin has. Provide evidence for your conclusion. What are the characteristics of this type of growth?

(B) **Describe** the recommendations that you might give to help the Florinese people. What policies (e.g., population, social, economic) should the government try to institute, bearing in mind the country's religious, economic, and social structure?

ANSWERS AND EXPLANATIONS

1. D

This question addresses your ability to recognize patterns of population growth from descriptions of the shapes of age-structure diagrams. Because you can discern information about population growth from age-structure diagrams, choice (E) is ruled out. The given description of an age-structure diagram is pyramid-shaped with a broad base. Zero growth would be indicated by an almost vertical (even slightly rectangular) age-structure diagram, which would rule out choice (A). Negative growth has a narrow base and broadens as you go up the pyramid, which rules out answer (B). Slow growth is characterized by a pyramid with a small base, so choice (C) is incorrect. Rapid growth is characterized by a pyramid with a broad base, so the correct choice is choice (D).

2. E

This question addresses your ability to identify phases of the logistical, S-shaped curve of nonhuman population growth. The lag phase is a short period in the beginning during which little or no growth occurs, so answer (A) is incorrect. The stable phase is a description of the phase where the carrying capacity or upper limit is reached; no growth occurs here, so choices (B) and (C) are incorrect. An oscillating phase can occur when the population has reached the carrying capacity, but again it is not a period of rapid growth, so choice (D) is incorrect. The log phase is another name for the period of rapid exponential growth, so choice (E) is correct.

3. C

This question assesses your ability to use the equation for changes in human population growth: $\Delta N_t = (B + I) - (D + E)$, when given data on rates of birth, death, immigration, and emigration. When you use the equation, countries (A) and (E) have net population increases of 5 people/1,000/yr., so choices (A) and (E) are incorrect. Countries (B) and (D) show net population increases of 15 people/1,000/yr., so choices (B) and (D) are incorrect. Only country (C) has no increase in population. In fact, its net population change is zero. So, choice (C) is correct.

4. A

This question assesses your knowledge of density-dependent and density-independent factors that limit population growth. Density-independent factors do not rely on how large or dense the population is or how they interact with other species or themselves. Environmental conditions such as pH, temperature, weather, climate, food supply, and water supply are all examples of density-independent factors. So, choices (B) through (E) are incorrect. Density-dependent factors do involve how large the population is and how it interacts with itself and other species. Factors such as predation and competition are density-dependent factors. So, choice (A) is correct.

5. B

This question assesses your knowledge of r-strategists versus K-strategists in nonhuman population growth. K-strategists have few offspring but nurture them for a long period of time. K-strategists include birds (e.g., eagles), mammals (e.g., whales, dogs), and some reptiles (e.g., alligators), so choices (A) and (C) through (E) are incorrect. In contrast, r-strategists have many offspring (often thousands) that develop and mature on their own with no parental care. Fish, invertebrates, and other insects (e.g., cockroaches) are examples of r-strategists. So choice (B) is correct.

6. D

This question assesses your ability to recognize the types of population policies and how they might be implemented in light of a country's social and religious beliefs. Pronatalist policies encourage high birth rates, high fertility rates, and large family sizes with the goal of increasing population growth. In contrast, antinatalist policies are those that encourage low birth rates, low fertility rates, and small family sizes with the goal of reducing population growth. While family planning is an important part of these policies, this country's social and religious beliefs were clearly stated to be against using contraception and abortion. Legally mandated sterilization, condoms, and birth control pills are all contraceptive methods, so choices (A), (B), and (E) would be incorrect. Tax breaks for large families would encourage people to have many children and would be against the goals of an antinatalist policy, so choice (C) is incorrect. Raising the legal age of marriage would reduce a woman's fertile period and would be consistent with the goals of an antinatalist policy, so choice (D) is correct.

7. E

This question assesses your ability to understand how age-structure diagrams can make predictions about population growth. After a long war when soldiers return home, there is usually a rise in births (e.g., post-war baby boom), so choice (B) is incorrect. This would result in a bulge, not a pinch, in the age-structure diagram, so choices (A) and (D) are incorrect. When plotted over several decades, the bulge in the age-structure diagrams would begin at the bottom and proceed upward as was the case in the post–WWII baby boom in the United States. So choice (C) is incorrect because the bulge was due to a boom, not a bust. Choice (E) is correct as the bulge is due to a boom and proceeds in the proper direction.

8. C

This question assesses your knowledge of the logistic equation for nonhuman population growth: $\frac{dN}{dt} = rN\left(1 - \frac{K}{N}\right)$. You seed the two flasks with the same number of algae, so the initial population size is the same. The rate of intrinsic growth (birth rates − death rates) would be expected to be the same between the two populations as they are the same species, so choices (A) and (E) are incorrect. Flask B has more of a vital nutrient, phosphate, which would allow it to have a higher carrying capacity (K). So, choices (B) and (D) are incorrect. Because the carrying capacity of Flask B is greater and the intrinsic growth rates are the same, according to the equation Flask B will have a higher rate of growth, so choice (C) is correct.

9. C

This question assesses your ability to recognize the consequences of habitat destruction in response to increasing human population growth. The city officials decided to fill in upstream wetlands to build new homes and accommodate the growing population. This policy had nothing to do with encouraging or discouraging birth rates and family sizes, as pronatalist and antinatalist policies do, so choices (A) and (B) are incorrect. Pollution and ecological stress would have nothing to do with flooding, so choices (D) and (E) are incorrect. The wetlands provided a buffer against floods in the river where water would spill into the wetlands rather than into the city. By destroying the wetlands for home sites, the excess water would have nowhere to go but to flood the town. This is an example of the consequences of habitat destruction, so choice (C) is correct.

10. A

This question assesses your ability to recognize characteristics of the phases of demographic transitions. The transitional phase from Pre-Industrial to Industrial phase is characterized by a drop in death rates (due to improved health care), while birth rates remain the same. Therefore, the population grows. Because the birth rates do not change, choices (B) through (E) are incorrect and only choice (A) has the appropriate changes listed. Therefore, choice (A) is correct

ANSWER TO FREE-RESPONSE QUESTION

To analyze the population growth of Florin, you must first construct an age-structure diagram from the available data. The shape of the diagram will allow you to make conclusions about the population growth of the country. You can then make recommendations about population policies, but you must also remember to consider the socioeconomic, political, and religious factors in the country. The ability to carry out any recommendations may well deal with the type of government.

(A) If you plot the data, you will get an age-structure diagram similar to the one below. You can characterize this diagram as a pyramid shape with a broad base. This allows you to conclude that **the population of Florin is a rapidly growing population**.

Characteristics of a rapidly growing population include high birth rates and high death rates (including high infant mortality). High death rates are also consistent with the lack of adequate health care as stated in the paragraph.

(B) Florin has a largely agricultural economy and most of the citizens are impoverished. The social and religious beliefs encourage large families with many children, who are essential contributors to the workforce. While the government wishes to raise the standard of living of the Florinese people, the **rapid population growth presents a problem**. Therefore, **you might suggest that the government implement an antinatalist population policy**. Following are three suggestions:

1. **Reduce birth rates**—This must be done without family planning policies, as the largely Catholic population would not support contraception or abortion policies; however, following are some ways of doing this without imposing family planning:

   - Raise the legal age of marriage so that women will be further into their childbearing years when married (i.e., shorter fertility periods).

   - Provide tax breaks to those families with fewer children (i.e., those with family sizes at or below the replacement levels).

   - While the totalitarian government may be able to mandate that women have fewer children, this might not be acceptable given the religious beliefs of the Florinese people. A better tactic would be to try to expand opportunities for women in the workforce. When women have more economic opportunities, they tend to have fewer children.

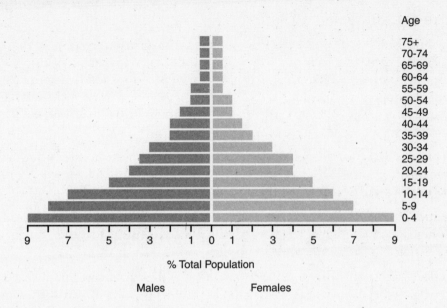

2. Try to break the poverty cycle

- Florin's main exports are wine and beer. They grow barley and grapes more to serve as cash crops than to provide food for the population. If the landowners could be enticed to reduce the acreage devoted to these cash crops and, instead, devote some acreage to food crops, more of the people could be fed.

- Foreign aid to Florin could involve direct food and medical care.

- Improved medical care would also reduce death rates, especially infant mortality. When the Florinese people see that their children survive longer, they will not need large families to replace lost children.

3. Institute education programs

- By educating women, you can expand their economic roles in society and hope to reduce birth rates.

- By educating women in prenatal and postnatal care (mainly nutrition), you can improve the survivability of infants and reduce infant mortality. Again, if people see that children are surviving, there will be less need to replace them.

CHAPTER 6:   HOW PEOPLE USE THE LAND

IF YOU LEARN ONLY FIVE THINGS IN THIS CHAPTER . . .

1. Several advances in agriculture have occurred (chemical fertilizers, chemical pesticides and herbicides, genetically bred or engineered crops, irrigation) to increase productivity, but these advances come at a cost (soil erosion and infertility, pollution, increased water demands). Sustainable agriculture applies ecological principles to farming and has less impact on the environment.

2. Forests, rangelands, and fisheries all hold renewable resources. However, the major dangers are taking too much from them (e.g., clear-cutting, overgrazing, overfishing) and pollution. These resources can be managed wisely.

3. Mineral resources are expensive to extract and process and are not renewable. Mining has devastating consequences on the environment in the form of pollution, soil erosion, and habitat loss. Conservation of mineral resources (e.g., reducing, reusing, recycling) along with rehabilitation of the mined lands are important in minimizing the impact of mining on the environment.

4. Public lands have many uses, such as forestry, mining, grazing, recreation, and nature preservation. Some of these uses are important economically but not necessarily environmentally friendly. There does not always have to be a conflict between economic and environmental interests, but both interests can work together for wise management of the country's resources without degrading the environment.

5. Ecological management of many resources in forests, rangelands, fisheries, mineral sites, and urban areas can be done wisely with the following considerations:

- Plan use of the land wisely—take into account economic, ecologic, social, and political considerations. Use sustainable land-use policies and take steps to minimize pollution and the degradation of the environment.

- Conserve the resources already available through reducing, reusing, and recycling—this minimizes the need for the resource.

- Take only what you need of the resource or what you can replace (renewable resources)—don't overgraze, extract too much of a mineral, overfish, or clear-cut trees.

- Preserve biodiversity to prevent diseases and pests and maintain the natural ecology.

- After use, rehabilitate the land through remediation and reclamation.

HUMAN NEEDS AND THE USE OF RESOURCES

Of all the species of life on Earth, human beings have a unique position. With our technologies, we are able to transform the planet, for good or bad. For example, our use of agriculture has greatly expanded our ability to feed ourselves, contributing to our population growth. We have fished the seas for food. We have cleared large areas of forests for both agriculture and living space. We have mined Earth's minerals for industry. As a consequence of these activities, we have placed stresses on Earth's ecosystems (pollution of air, water, and ground, and loss of biodiversity). In this chapter, we will explore how humans use Earth's resources and the resulting consequences.

AGRICULTURE: HOW TO FEED A GROWING POPULATION

Humans need a diverse, balanced diet consisting of carbohydrates, fats, and proteins that are essential for providing and storing energy. Proteins are also broken down to amino acids that are important structural components of cells. According to the U.S. Department of Agriculture, the daily nutritional requirement for a human is as follows:

	Grams	% kcal (total intake = 2,000 kcal)
Protein	91	18
Carbohydrate	271	55
Total Fat	65	29

Source: Courtesy of USDA, available at www.health.gov/
dietaryguidelines/dga2005/document/html/chapter2.htm#table2.

We often see labels that denote Calories with a capital C. One Calorie is equal to 1,000 calories (lower case) or 1 kcal. So, a 2,000 Calorie diet is actually 2,000 kcal or 2 million calories.

In addition to these main nutrients, we must take in essential vitamins and minerals (those that our body cannot make itself). Vitamins and minerals are important cofactors for enzymes that conduct the biochemical reactions in all the cells of our bodies. According to the USDA (www.health.gov/dietaryguidelines/dga2005/document/html/chapter2.htm#table2), daily requirements of these are as follows:

- Sodium—1.78 g
- Potassium—4.04 g
- Calcium—1.32 g
- Phosphorus—1.74 g
- Magnesium—380 mg
- Copper—1.5 mg
- Iron—18 mg
- Zinc—14 mg
- Thiamin—2.0 mg
- Riboflavin—2.8 mg
- Niacin—22 mg
- Vitamin C—155 mg
- Vitamin B6—2.4 mg
- Vitamin B12—8.3 mg
- Vitamin E—9.5 mg
- Vitamin A—1.05 mg

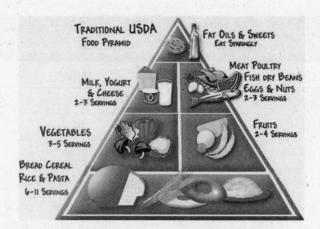

Figure 6.1 The 1992–2000 Food Pyramid
Source: Courtesy of USDA and DHHS.

To achieve these nutritional requirements, we must eat balanced diets of meat, fruits, vegetables, grains, and dairy products. In 1992, The U.S. Department of Agriculture (USDA) and the Department of Health and Human Services (DHHS) created the Food Pyramid. (See Figure 6.1.) This graphic provided a daily guide to servings that a person should have for a healthy diet. The USDA revised this pyramid to reflect new nutritional guidelines (2005 Dietary Guidelines for Americans) released in January 2005. The new pyramid is interactive to correlate an individual's lifestyle (age, sex, physical activity) to proper nutritional food choices (see www.mypyramid.gov for details).

Although these are recommended dietary guidelines, most of the world's population does not consume such a diet. According to the Food and Agriculture Organization of the United Nations, the number of undernourished people in less developed countries (LDCs) increased during the 1990s to over 826 million people, approximately 13.7 percent of the world population. People in more developed countries eat more meat than those in LDCs. People in LDCs rely primarily on grains in their diet. Because of this, advancements in agriculture will be important in relieving world hunger.

TYPES OF AGRICULTURE

Agricultural technology has existed for almost 10,000 years. The basic unit of agriculture is the farm, where farmers clear the land, plant seeds, grow crops, and harvest them. Farmers must carefully manage their resources, such as land, soil, water, and seeds. They must care for the growing crops to keep them free from weeds and pests. The harvested crops can be used to feed the farmer's family (subsistence farming), to feed livestock, or be sold for profit (small business, large commercial farming).

Farms can vary in terms of who owns and works the land:

- **Small tenant farms**—The farmers do not own the land, but rather, they work the land for a landlord. The farmers may receive pay or a share of the crops in return for their work.

- **Family-owned farms**—A family owns the land and works the farm for itself (subsistence farming, small business farming). Family-owned farms can be small or intermediate in size.

- **Commercial farms**—Agribusiness companies own large farms and employ farmers to work for them. The crops are sold or used for other products (e.g., corn oil, high fructose corn syrup).

- **State-owned farms**—The government owns these large farms and employs citizen farmers to work the land. State-owned farms were largely found in the former Soviet Union and China.

Ownership varies depending on the countries in which the farms are located. In the United States, individuals and/or families own the majority of farms (approximately 60 percent). Worldwide, tenants work approximately 40 percent of farms.

Regardless of who owns the farms, there are two types of agriculture (see Table 6.1):

1. Conventional, or unsustainable agriculture

2. Agroecology, or sustainable agriculture

Table 6.1 Comparison of Conventional and Sustainable Farming Methods

Parameter	Conventional	Sustainable
Altering land	Till fields, irrigate	Maintain natural environment
Number of crops/ field	One	Many crops together
Harvest	All at once	Different crops have different harvest times
After harvest	Plot is bare	Plot is never bare
Susceptible to soil erosion	Yes	No
Herbicide use	Yes	No
Pesticide use	Yes	No
Fertilizer use	Yes	No

In conventional agriculture, farmers strip the field of its natural environment by removing trees and stumps, tilling the soil, and/or slashing and burning (vegetation is cut down and burned to expose the soil). Farmers prepare the fields by plowing; in some cases, the fields must first be irrigated to bring in adequate water. Farmers prepare the soil with fertilizers and plant one seed crop per plot. The farmers tend the fields as the crop grows and treat the crops with herbicides to remove weeds and pesticides to remove insects. When the crop is ready, farmers harvest the crop, thereby leaving the field bare. Later, the field must be plowed again.

Conventional agriculture disrupts the natural ecology of the field in at least five ways:

1. The field's natural ecology gets upset when the land gets stripped in preparation for farming. Displaced herbivores without natural vegetation then feed on the planted crops.

2. Chemical fertilizers, herbicides, and pesticides leach into the soil and groundwater. Rains then wash these chemicals into nearby streams and rivers. For example, rains wash phosphate-rich fertilizers into local streams. The phosphates stimulate algal blooms. The thick layers of growing algae mature and die quickly. Their decomposition uses up the oxygen in the water, thereby suffocating fish and causing fish kills.

3. Single crops tend to deplete the soil of vital nutrients, which must then be replaced artificially. Alternatively, the fields must be plowed and allowed to lie fallow every few years to regain vital nutrients.

4. After harvesting, the bare fields are susceptible to soil erosion by wind and rain.

5. Use of commercial seeds, fertilizers, herbicides, and pesticides are expensive and constitute a large portion of the farm's operating costs.

Unlike conventional agriculture, sustainable agriculture applies ecological principles to farming. Farmers plant multiple crops together in the same fields (e.g., alternating rows of corn, soybeans, and squash in the same field); ideally, the crops should have different harvest times. Farmers use natural fertilizers such as compost and manure. They also seed plants that are natural insect repellents or ground covers to protect their crops. As each crop matures, they harvest it; where possible, they replant the seeds from the harvested crops.

In contrast to conventional agriculture, six practices of sustainable agriculture help to avoid disrupting the land:

1. By not clearing the fields entirely, natural ecosystems of herbivores are maintained with the natural vegetation. Therefore, they do not eat as much of the planted crops. Furthermore, predators keep the herbivore population in check.

2. By planting many crops in the same fields with various harvest times, the land is never bare and subject to soil erosion. Herbivores do not eat all of one crop but rather some of each crop. Therefore, an entire crop is not destroyed.

3. Planting diverse crops helps to build or retain disease resistance among the crops, retain moisture in the soil, and supply nutrients to the soil, thereby maintaining fertility.

4. The use of natural fertilizers, ground cover to reduce weeds, and insect-repellent plants rather than chemicals prevents pollution of the surrounding land and water.

5. Because the agricultural system sustains itself, the farmers spend less energy tending the fields.

6. The use of natural components for crop seeds, fertilizer, pest control, and weed control reduces the operating expense of the farm.

While sustainable agriculture is ecologically better, it does not yield as many crops as conventional agriculture does. Therefore, products from sustainable agriculture tend to be more expensive in the marketplace than those produced by conventional agriculture.

Conventional agricultural methods do not always yield high productivity. Much about the productivity depends upon the country and economic status of the farmer. In the United States and many MDCs, farmers can afford the seeds and chemicals used to make the land highly productive (fertilizers, pesticides, herbicides, etc.). However, many poor farmers in LDCs cannot afford these resources to increase productivity.

GREEN REVOLUTION

The high productivity of conventional agriculture methods is a direct result of the so-called **Green Revolution**. The Green Revolution was a cooperative venture between the Western nations to help farmers in India and Mexico increase their productivity and relieve hunger in these areas. The Green Revolution began in the 1940s with research funded by the Rockefeller Foundation to improve agricultural yields in Mexico. A plant scientist named Norman Borlaug developed high-yield varieties of wheat that were more resistant to pests and diseases than other varieties. These new wheat plants were shorter, had more numerous and stronger stalks, and held more heads of grain. When used with irrigation and chemicals (fertilizers, pesticides, herbicides), these new wheat plants increased the production of grain per area and raised the income of the Mexican farmers; however, despite these advances, many Mexican farmers are still poor.

In the 1960s, scientists developed new high-yield varieties of other crops such as rice, sorghum, corn, and beans. These new varieties grew quickly, so that farmers could plant and harvest crops faster and more often. The intensive farming of the land helped relieve famine in Latin America and Asia.

However, the agricultural technologies developed during the Green Revolution have not been successful in all areas of the world. For example, in sub-Saharan Africa where the climate is arid and irrigation is not as feasible, the Green Revolution technologies have not been nearly as successful as they were in Latin America and Asia. Also, farmers in Africa do not have access to pesticides and fertilizers as farmers in other countries do. Therefore, new drought-resistant varieties of crop plants must be developed for use in this area.

Although Green Revolution technologies have been successful in some areas, they make the farmers dependent upon chemicals and irrigation, which can have negative impacts on the environment. However, if properly managed, they increase the productivity of farmland, thereby reducing the amount of land needed for farming.

GENETIC ENGINEERING AND CROP PRODUCTION

The Green Revolution relied on creating new varieties of high-yield crops by using genetic crossbreeding techniques. In this method, scientists cross-pollinate plants with desired traits (insect resistance, drought resistance, etc.), grow the new plants from seed, and test them for desired characteristics. This process is long and labor-intensive because scientists may have to go through many generations of crossbreeds to obtain crops with the desired traits.

Since the Green Revolution, biotechnology has provided new methods of producing crops with desired characteristics. If the scientists can isolate a gene from one source (plant, bacteria) for a desired characteristic (e.g., insect resistance), they can make thousands of copies of the gene, and place those genes in another plant. The receiving plant can take up the DNA, express the foreign gene, and even pass that gene on to its offspring. This process is called **genetic engineering** and is more efficient than conventional genetic approaches.

Let's look at an example of genetic engineering. Tomatoes, and other fruits, produce a gas called **ethylene**, which makes them ripen quickly. One problem with transporting tomatoes is that if they are picked while ripe, they will spoil on the way to market. So, often tomatoes are picked and shipped while green. At the market, the grocers spray the fruit with ethylene to start the ripening process. However, it is believed that this procedure alters the flavor of the tomatoes compared to those that are naturally ripened on the vine. If you could genetically alter the tomato plant to inhibit ripening, you could leave the tomatoes on the vine longer and increase the shelf-life once they went to market.

To address this problem, a company called DNA Plant Technologies genetically engineered (GE) a new tomato plant. An enzyme called ACC synthase produces ethylene, which is the major ingredient in the ripening process. Scientists placed a nonsense copy of the ACC synthase gene into tomato plants. The nonsense copy does not get translated to make ACC synthase, so ACC synthase levels go down and ethylene does not get produced. The GE tomato, which was called Endless Summer (see Figure 6.2), did not ripen as fast and had a much longer shelf life. While the GE tomatoes were briefly tested in the marketplace, patent arguments with another company prevented DNA Plant Technologies from fully marketing their product and the Endless Summer tomato was discontinued.

Figure 6.2 Genetically Engineered Tomato
Source: Photo courtesy of USDA. (Credit: Jack Dykinga)

To date, GE crops have included the following modifications:

- **Insect resistance**—A bacterial gene (Bt) has been placed into corn, cotton, potato, and tomato plants.

- **Herbicide resistance**—Genes from various sources have been incorporated into corn, soybeans, cotton, rice, and other plants.

- **Virus resistance**—Plant virus genes have been placed in squash, potatoes, and zucchini.

- **Delayed fruit ripening**—Bacterial and viral genes have been placed in tomatoes, as discussed above.

- **Altered oil content**—Altered soybean genes have been placed in soybean and canola plants to change their oil content.

- **Pollen control**—A bacterial gene has been incorporated into corn plants to control their pollen production.

While GE crops hold many promises for food production, the following three drawbacks exist that have mainly to do with public perception:

1. When grown in the wild, genes from GE crops might pass to native crops and upset the ecological balance.

2. Pests and weeds could evolve resistance to the GE genes, thereby reducing the effectiveness of the GE crops and making the situation worse.

3. When making GE plants, some genes that contain allergens might be transferred as well (e.g., peanut allergens).

While there has been little evidence of these problems in laboratory tests and limited field tests, consumer groups and consumers have shown great resistance to using and marketing GE crops, especially in Europe. This lack of consumer confidence has spread to other countries, such as the United States, Japan, and Australia, and has caused governments to proceed cautiously in approving GE foods for market.

DEFORESTATION AND IRRIGATION

To obtain land for agriculture, farmers must clear large tracts of forested land. This clearing of forests, especially tropical rain forests, is called **deforestation**, the process during which trees are cut down and/or burned. Deforestation mostly occurs for agricultural purposes, though it is also done for logging. (See Figure 6.3.)

In LDCs in or near the tropics, farmers typically clear the land by slash and burn techniques. The clearing of the land usually provides a few years of intensive farming. However, in tropical areas, most of the nutrients lie not in the soil but in the biomass of the trees. By removing the trees, most of the nutrients are lost and the land quickly becomes infertile. Furthermore, loss of trees facilitates erosion of the soil, increasing the risk of deadly landslides, as well as providing new breeding grounds for mosquitoes and other pests. Farmers must then move on to other areas of the forest for new agricultural plots. In the 1980s and 1990s, as much as 16 percent of the Amazon rain forest was chopped down. The rates of deforestation vary from region to region.

Deforestation has at least three negative consequences:

1. Tropical rain forests take up much of the carbon dioxide in the atmosphere, a greenhouse gas that absorbs heat and plays a crucial role in global temperatures. Deforestation causes carbon dioxide to stay in the atmosphere and contributes to global warming.

2. Evapotranspiration through trees returns precipitated water to the atmosphere as part of the water cycle. Deforestation reduces this process and disrupts the water cycle.

3. Tropical rain forests are the most biodiverse ecosystems in the world. Deforestation eliminates this ecosystem and the species that live there.

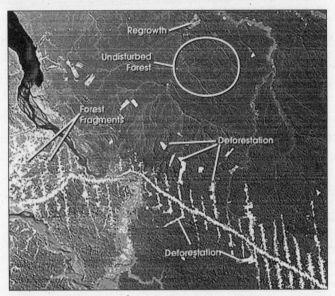

Figure 6.3 Deforestation
This satellite image of the Brazilian state of Para in the
Amazon region was taken in 1986. The dark areas are
forest, the white areas are deforested areas, and the
gray areas are regrowth areas. Note the "fishbone"
pattern of deforestation that occurs along roads.
Source: Photo courtesy NASA, available at http://earthobservatory.nasa.
gov/Library/Deforestation/Images/figure2.jpg.

A recent study by NASA and Duke University projected that deforestation in the Amazon, Central
Africa, and Southeast Asia would affect precipitation patterns far from the local areas. Precipitation
patterns in North America could be affected by deforestation in these areas.

Irrigation is an important method of bringing water to agricultural crops in many areas of the
world and was one major factor in the Green Revolution. Irrigation uses about 60 percent of the
world's freshwater supplies. Irrigation techniques include the following:

- **Flood or furrow irrigation**—Water gets pumped into trenches that are dug in the fields
  and flows along the ground to the crops. Though this method is simple and cheap, over half
  the water never reaches the crops.

- **Drip irrigation**—Water flows through pipes with holes in them. The pipes are laid
  alongside the crops, allowing the water to sustain them.

- **Spray irrigation**—Long hoses with spray attachments are placed in the fields, much like
  watering your lawn at home. This system requires machinery for pumping and spraying.

The water used in each of these techniques mostly comes from surface waters, but it can also come
from groundwater.

Increased agricultural demand has created several environmental problems. First, the supplies of surface freshwater and groundwater available for drinking become reduced as agricultural demands increase. Second, irrigation runoff carries chemicals (fertilizers, pesticides) and soils into local water sources, thereby polluting them. Third, removing groundwater from aquifers makes them saltier and unusable for drinking water. Fourth, irrigation projects (e.g., dams) alter the natural flow of water through ecosystems, thereby depriving them of water. Finally, irrigation and constant evaporation makes topsoil very salty as well. Therefore, efficient irrigation systems and water conservation methods will be needed to keep agricultural productivity high, while preserving and protecting valuable drinking water resources.

CONTROLLING PESTS

Pests, most notably insects, worms, and rodents, have long damaged agricultural crops. Some agricultural crops tend to be more susceptible to pests than others. Native plants, for instance, have typically evolved resistance to native pests, but non-native monoculture crops, common in industrial farming, have not, and require heavy treatments to minimize pest damage. Technology has enabled farmers to control pests with chemicals. The major types of chemical pesticides include the following:

- **Organochlorine insecticides** (DDT, chlordane)—cause convulsions, paralysis, and death. They accumulate in the food chain and their use has been banned in the United States for health and environmental concerns, but they are still used in LDCs.

- **Organophosphate pesticides** (malathion, parathion)—disrupt an enzyme that regulates acetylcholine, a neurotransmitter. Although they break down rapidly in the environment and do not accumulate in food chains, they can contaminate water sources and are highly toxic to humans. Despite these dangers, organophosphate pesticides are still used, especially for mosquito control in wetland, swampy areas of the United States.

- **Carbamate pesticides** (carbaryl, propoxur)—also interfere with acetylcholine metabolism. They also do not accumulate in food chains, and their toxicity varies.

- **Pyrethroid pesticides**—artificial versions of pyrethrin, a naturally occurring pesticide in chrysanthemums, and toxic to the nervous system. While they break down rapidly in the environment and do not accumulate in food chains, they are expensive.

- **Microbial pesticides**—derived from bacteria (Bt) or other microorganisms. They can be made specifically for certain pests, do not accumulate in food chains, do not last long in the environment, and have low toxicity.

- **Biochemical pesticides**—naturally occurring chemicals that do not necessarily kill the insects. Instead, they may interfere with mating or attract insects into traps. They are not toxic.

In the past, farmers heavily used pesticides to control multiple types of insects. However, a new philosophy called **integrated pest management** (IPM) has recently come into favor. IPM is a series of pest management decisions and actions that take the following four questions into consideration:

1. What level of pests can I tolerate before I must take action? Just because you see some pests does not mean that they are a problem.

2. What types of pests are prevalent in my fields? Some pests are harmless; others are not. If you know which pests you must prevent, you can make reasonable decisions.

3. How can I prevent pests? Rotating crops or planting pest-resistant crops may prevent pests from becoming a problem.

4. How can I control pests? If you can monitor and identify the pests, you can select appropriate control measures, such as using biochemical pesticides or trapping pests. If necessary, limited use of broad pesticides may be in order.

IPM is economical, ecological, and designed to be the least damaging to health and the environment.

In the United States, the Environmental Protection Agency (EPA) regulates the use of pesticides by licensing who can sell or distribute pesticides, evaluating the potential health and environmental risks, registering pesticides, and setting limits on how much can be used. While an optimal goal would be for a pesticide or other food additive to have a zero chance of causing cancer in humans, for practical reasons the EPA allows a substance to be present in food if it has no more than a 1 in 1,000,000 chance of causing cancer. The EPA also provides information for training workers in effective use of pesticides and safety information for consumers on reducing the risks of pesticides.

FORESTRY AND RANGELANDS

Much of the United States is covered by forests and rangelands. These areas provide many national resources and must be managed carefully.

FORESTS

A *forest* is traditionally defined as land where at least 10 percent has been or is currently covered by trees and is not built up or used for agriculture. Forests cover one-third of the United States (see Figure 6.4) and include both commercial and noncommercial forests. Commercial forests provide timber for construction, fuel, and paper. Noncommercial forests are parks, wildlife refuges, and wilderness areas.

Forest management is different from agricultural management in that forests take much longer to regrow than agricultural crops do. Cropland can be regrown annually, but forests may take 20 to 1,000 years to regrow. Thus, forestry management involves a cycle of decisions between planting and harvesting called a **rotation**. The steps in a rotation involve the following:

1. Taking inventory—What types of trees are on a site? What are their ages, numbers, and distributions?

2. Developing a forest management plan on how to manage the site

3. Building roads for access

4. Preparing the site

5. Harvesting the timber

6. Regrowing and managing the site for the next harvest

Tree harvesting methods consist of the following:

- **Selective cutting**—Only intermediate-aged or mature-aged trees are cut singly or in small groups. This method preserves biodiversity by reducing crowding and fostering the growth of younger trees. It also protects the site from erosion and fire damage. However, it is expensive.

- **Shelterwood cutting**—This system removes mature trees in several cuttings over a long period (about 10 years). This method allows younger trees to grow, maintains good wildlife habitats, and eventually converts an uneven-aged stand into an even-aged stand (*stand* is an area of trees within a forest). It also prevents erosion and allows the trees to seed naturally.

- **Seed-tree cutting**—Nearly all of the trees are cut except a few mature, seed-producing trees in an even distribution. The seed-producing trees reseed the area naturally and may be cut down later when younger trees become established. This method helps prevent erosion while maintaining a good forest for recreation and wildlife conservation.

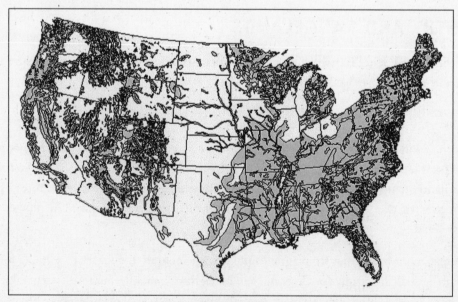

Figure 6.4 Distribution of Forests Across the United States

Source: Courtesy of USEPA, available at www.epa.gov/emap/images/data/forest.gif.

- **Clear-cutting**—This practice removes all of the trees in an area in one single cut. The area is then reseeded naturally or artificially. This is the most common method of tree harvesting because it is inexpensive. Roughly two-thirds of the annual timber harvest is done in this way. However, this method destroys habitats, reduces biodiversity, and leaves large patches of land bare and vulnerable to erosion. Following are types of clear-cutting now practiced:

  — *Whole-tree harvesting*—a clear-cutting method in which all of the trees are cut at the base and all wood debris is removed. It deprives the soil of nutrients and destroys habitats.

  — *Strip logging*—a clear-cutting method that removes small strips from the site in several cuttings over several years rather than removing everything from a large area at once. This type of clear-cutting minimizes habitat damage and erosion.

Forestry management consists of two major approaches: even-aged and uneven-aged. **Even-aged management** involves a site in which the trees are about the same size and age. In even-aged management, foresters and loggers clear-cut the site of all or most of the trees and then replant them, either naturally or artificially. When the trees reach maturity, the area is harvested and replanted again. This approach is useful in tree farms that grow one type of tree (monoculture). **Uneven-aged management** involves a site where the trees are different sizes and ages. In uneven-aged management, the goals are to maintain natural forest regeneration, increase biodiversity, and maintain long-term production of timber for multiple uses. Uneven-aged management utilizes selective cutting methods and clear-cutting of only small areas where those tree species benefit by it.

Tree farms or **plantations** use the even-aged management approach to manage forests and produce forest products (timber, fuel). They involve clear-cutting an area, planting one type of tree, managing the area (sometimes with pesticide and fertilizer use), and harvesting the products. While this method appears to be a good one, growing only one type of tree reduces biodiversity and makes the plantation more susceptible to pests and diseases. Furthermore, clear-cutting methods can increase soil erosion.

Environmentalists and foresters are striving for sustainable forests with the following techniques:

- Recycling paper—reducing demand for timber

- Using long rotations—100 to 200 years

- Using selective cutting—preserving biodiversity

- Protecting/minimizing soil erosion

- Strip logging instead of general clear-cutting

- Leaving dead trees, fallen timber, and wood debris—maintaining habitats, recycling nutrients, and restoring soil fertility

- Relying on natural mechanisms (biodiversity) of pest and disease control—biological controls, integrated pest management

A more controversial aspect of forest management is how to handle forest fires. Forest fires can start naturally, usually by lightning, or be started by humans (e.g., from not extinguishing campfires or from burning trash). For some tree species, such as the giant sequoia and jack pine, fires are an important part of the life cycle and activate seed germination. Therefore, the U.S. Forest and Wildlife Service has adopted a controversial policy that allows wildfires to burn in national parks and wilderness areas. However, as droughts increase, the occurrences and severity of wildfires also increase and could potentially destroy large areas of forest. The increasing number of homes adjoining forest areas also puts property and lives at risk from wildfires, further complicating fire management strategies. Protecting forests from damaging fires involves the following:

- **Prevention**—This involves educating people, issuing burning permits, and closing areas of forests to recreational users in times of drought.

- **Prescribed burning**—Foresters burn selected areas in a controlled manner for a limited time to prevent leaf litter buildup and reduce outbreaks of pests and diseases.

- **Presuppression**—This method involves detecting and controlling a fire at the early stages to reduce its spread and damage. Firefighters use firebreaks and fire roads to clear vegetation and prevent the spread of wildfires.

- **Suppression**—Firefighting of major wildfires uses large-scale firebreaks, water, fire-retardant chemicals, and backfires to contain the affected areas.

National forests in the United States provide major sales for the timber industry, and supplements to the Forest Service budget come from these sales. However, by law, the U.S. government cannot profit from timber sales, and that means the timber trees are harvested and sold to the timber industry well below their market value. This practice basically means that the U.S. government is subsidizing the timber industry, resulting in increased harvesting from public lands. To better manage national forests and reduce these practices, environmentalists would like to see the following occur:

- Ban timber cutting in national forests

- Eliminate new road building in national forests

- Require that timber sales from national forests bring a profit to the taxpayer based on market value

- Do not supplement the forest service budget through timber sales

RANGELANDS

Rangelands are areas of land that supply forage or vegetation for grazing (grass-eating) and browsing (shrub-eating) animals but are not managed. Rangelands include prairies, desert scrub, grasslands, chaparral (shrublands), open woodlands, riparian (areas alongside streams and rivers), and arctic tundra/desert. Worldwide, about 40 percent of rangelands are devoted to livestock. About 25 percent of the United States is rangelands (see Figure 6.5), and while most rangeland is

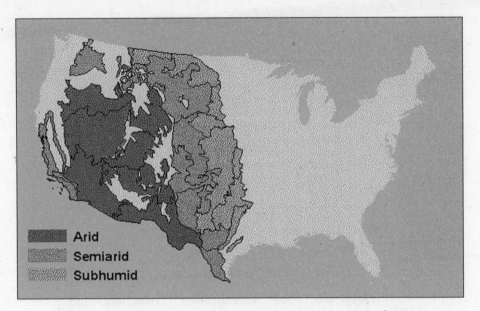

Figure 6.5 Distribution of Rangelands Across the United States
Source: Courtesy of USEPA, available at www.epa.gov/emap/images/data/arid.gif.

privately owned, extensive grazing does take place legally on public lands. Rangelands tend to be in areas of low to moderate rainfall (arid, semi-arid).

The major environmental problem in rangelands is the **overgrazing** that occurs when too many animals graze for too long on a given area, exceeding the carrying capacity of the rangeland. Overgrazing removes vegetation and compacts the soil so that it no longer holds water, thereby hastening soil erosion. Overgrazing can also lead to desertification and loss of biodiversity.

The guiding principle in managing rangelands is to raise the largest number of animals that a given patch of rangeland can support without overgrazing. To do this, ranchers must control the **stocking rate**, that is, the number of a particular animal on a given area that does not exceed the carrying capacity. In addition to the stocking rate, ranchers must also pay particular attention to the distribution of animals in a given area. If left to themselves, many grazing animals will congregate around water. They eat and overgraze the vegetation in these areas, thereby depleting the biodiversity and allowing invasive species or undesirable grasses to gain a foothold.

The most common methods of grazing are as follows:

- **Continuous grazing**—year-long or season-long grazing in a particular area. This is the type that is most commonly used in the United States. It is less costly and requires little in the way of fencing or animal handling.

- **Deferred grazing**—moving livestock between two or more grazing areas. This method is more labor-intensive, but it allows grasses to recover from grazing. Overall, this method is better for protecting the quality of the rangeland.

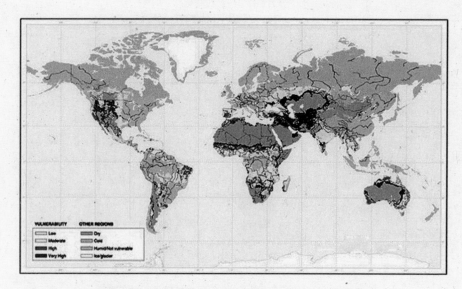

Figure 6.6 Global Desertification Vulnerability

Source: Courtesy of USDA, available at http://soils.usda.gov/use/worldsoils/mapindex/desert.jpg.

Other ways of managing rangelands include using herbicides to suppress unwanted vegetation, seeds and fertilizers to increase desired vegetation, and traps/hunting to control predators.

Sustainable management of rangelands tends to reduce the stress on the land, allow time for recovery, and preserve biodiversity. Following are four methods for sustainable rangeland management:

1. Issuing permits to limit the number of ranchers that can use public lands for grazing

2. Reducing the number of livestock on grazing lands

3. Erecting fences around streams and rivers to preserve these habitats, while increasing funding to restore these areas

4. Preventing grazing on poorly conditioned rangelands until they can recover

Besides overgrazing, rangelands are highly sensitive to drought conditions. As most rangelands are already semi-arid and arid lands, any decrease in rainfall can be devastating to the condition of the rangeland. Drought, along with overgrazing and poor management, can reduce vegetation, cause erosion of topsoil, reduce water supplies, and leave the land vulnerable to desertification. The global desertification vulnerability map in Figure 6.6 is based on soil climate and soil classification.

MINING AND FISHING

The United States has rich mineral and fishing resources. However, overmining and overfishing have damaged the land and waters. These resources must be carefully managed to sustain them and maintain the integrity of the environment.

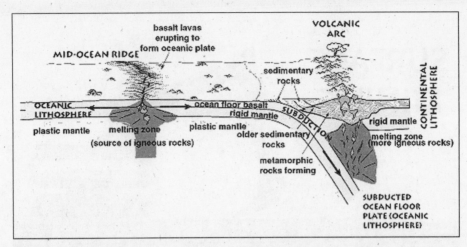

Figure 6.7 Sketch of Plate Tectonics and Formations of Various Rock Types
Source: Courtesy of USGS.

MINING

Minerals are nonliving natural substances with limited chemical composition and an orderly atomic arrangement. They can include the following substances:

- **Precious metals**—gold, silver, copper, aluminum, cobalt, platinum, titanium, iron, etc.

- **Nutrients**—phosphorus, nitrogen compounds

- **Energy sources**—coal, oil, natural gas, uranium

Minerals were present when Earth was formed, while fossil fuels formed from the remnants of living organisms long ago. Minerals have been transformed over time through geological processes (intense heat, pressure). The process of mineral formation is closely linked to plate tectonics. (See Figure 6.7.) Hot magma in Earth's mantle dissolves minerals. When magma comes to the surface, it cools and various minerals crystallize out of solution to form mineral deposits in the crust. Magma comes to the surface in volcanically active areas of seafloor spreading (mid-oceanic ridges), volcanic hotspots (e.g., Hawaiian Islands), and subduction zones (where one plate goes under another). Furthermore, mineral deposits may be brought to the surface by mountain-building processes (folding, faulting). Erosion of overlying rocks by wind and water may also expose mineral deposits.

Most mineral deposits must be extracted through two basic mining techniques:

1. **Surface mining** is used for mineral deposits near the surface. (See Figure 6.8.) The overlying vegetation, soil, and rocks, called **overburden**, must be removed to expose the ore deposit. In the United States, about 90 percent of ores are extracted by surface mining techniques, such as the following:

Figure 6.8 Surface Mining for Coal
Source: Courtesy of USDOE.

- *Open pit surface mining*—Miners dig a large pit or quarry and remove the exposed ore. This technique is used to extract copper, iron, sand, stone, gravel, limestone, granite, and marble.

- *Area strip mining*—This technique is used on flat or rolling hill terrains. Miners dig a series of parallel trenches to expose the ore. After the first trench is dug and the ore is extracted, it is used to store debris and overburden from the second trench. When the ore is extracted from the second trench, it is used to store debris and overburden from the third trench, and so on. When all of the ore is extracted, the land may be reclaimed or just left as a series of mounds of debris that then erode away.

- *Contour strip mining*—This technique is used on steep hills or mountainous terrains. Miners dig a series of terraces into the hillside starting from the bottom and extract the ore. The overburden from each upper terrace gets stored in the next lower terrace. When all of the ore gets extracted, the land can be reclaimed or left as a steep series of banks of rock and soil that are susceptible to erosion. If bulldozers cannot economically remove overburden, then miners can drill holes into the hillside to expose and extract the ore.

- *Mountaintop removal*—This is a controversial type of surface mining in which the entire summit of a mountain or mountain ridge is cut away to expose a coal seam beneath. After the seam is mined, the overburden is replaced on the summit and any excess is dumped in nearby valleys. Despite replacing the overburden, the original contour of the mountain cannot be fully restored.

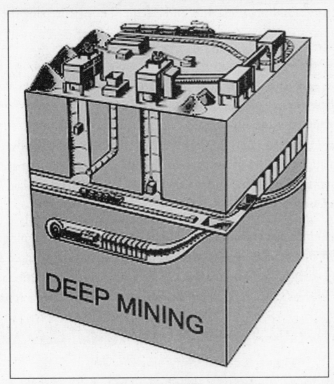

Figure 6.9 Deep Mining
Source: Courtesy of USDOE.

2. **Deep mining** is used for mineral deposits far below the surface. (See
   Figure 6.9.) Miners must dig deep shafts and tunnels into the ore seam.
   They must then use elevators to make these tunnels accessible to miners and
   machinery. The tunnels must be well-ventilated and structurally supported.

Mining disrupts the environment. Surface mining degrades topsoil and leaves large
segments of the land vulnerable to erosion. In both types of mining, miners pile
discarded rock called **tailings**. These tailings contain hazardous substances such as
lead and mercury. As water pours through the tailings and the mine itself, it picks
up these toxic substances. The contaminated water makes its way into surface
streams and rivers as well as groundwater, thereby contaminating these sources.
Waste products from mining represent the single largest source of solid waste in
industrialized countries.

Once extracted, the minerals in the ore must be processed to separate the
desired ore from the undesired surrounding rocks. The mined ore gets crushed,
concentrated, and physically or chemically treated to refine the mineral in the ore.
Waste impurities get disposed in the tailings. Tailings from processing are usually
fine-grained particles that can become airborne, and they typically contain toxic

AP EXPERT TIP

Water mixes with the sulfur
in the tailings and creates
sulfuric acid. This acid
runs off into waterways
contributing to acid mine
drainage (AMD). AMD
causes the pH levels of
waterways to drop to
levels where fish and other
aquatic organisms can no
longer reproduce.

substances from the rock such as cadmium, lead, zinc, and arsenic. Thus, processing ore can cause soil, water, and air pollution.

Several government agencies—the Department of Interior's Office of Surface Mining and Bureau of Land Management—regulate mining activities in the United States. States also impose regulations on mining activities. These regulations deal with how mining can be done and how the environment should be treated during and after mining operations. For example, the Surface Mining Control and Reclamation Act of 1977 requires mining companies to replant vegetation on land that was strip-mined. In addition to environmental concerns, the federal Occupational Safety and Health Administration as well as the Mine Safety and Health Administration make regulations regarding the exposure and safety of miners.

Perhaps the best way to reduce the environmental impact of mining is to conserve mineral resources. Reusing and recycling metal objects like aluminum cans will reduce the need for mining new resources, while reducing energy consumption reduces the need for coal. Additionally, governments must protect wild areas from mineral exploration and mining to force conservation measures.

FISHING

Worldwide, people get approximately 20 percent of their protein from fish and shellfish. Most of the fish in our diets comes from the estuaries and open oceans within 200 miles of the coasts. Commercial fishing for profit involves large fleets of ships that go out in the ocean to catch fish and shellfish that are then sold to processors and consumers for profit. The major types of fish and shellfish caught commercially include the following:

- **Bottom-dwelling fish** (demersal)—cod, haddock, flounder
- **Open water fish** (pelagic)—tuna, mackerel, sardine, herring, salmon, anchovy
- **Crustaceans**—shrimp, crab, lobster, krill
- **Mollusks**—oyster, clam, scallop, octopus, squid

Commercial fishing vessels use advanced sonar methods to locate schools of fish. Sometimes, they use lights and sound devices to lure fish toward them. Once the fish are located, the following methods are used to harvest them and these methods utilize nets, hooks, and traps:

- **Purse-seine fishing**—One or more boats deploys a large net around and underneath a school of fish; the net looks like a drawstring purse. Fishers pull the net in and retrieve the fish. This method is used for pelagic fish such as tuna.
- **Trawling**—A fishing boat deploys a large funnel-shaped net that is held open by two wooden flaps. The boat moves along in the water and scoops fish into the net. Fishers pull the net up and dump the bag on the deck of the ship to retrieve the fish inside. This method is used for demersal fish, such as cod and flounder. Shrimp are also harvested in this manner.

- **Drift net fishing**—A fishing boat deploys a long ribbon-like net kept afloat in the water by buoys. The net forms a long fence in the water and is allowed to float freely for several hours or days. As fish swim into the net, their gills get caught in the spaces of the net. The fishers reel in the net and harvest the captured fish. This method is used for pelagic fish.

- **Hook and line**—This method includes multiple techniques, including trolls, handlines, and rod and reel. A boat places multiple short lines with baited hooks attached in the water. The lines may be stationary or the boat may move them through the water. Fish attack the bait and become hooked. The fishers reel in the lines to harvest the fish. This relatively sustainable method is most commonly used in recreational fisheries. A special type of hook called a jig is used to snag squid.

- **Pole and line (bait boat fishing)**—This is a type of hook and line fishing in which some schooling fish such as tuna are attracted to the surface and hooked singly with fishing poles and baited hooks. The fishers reel the fish out of the water and into the boat. Some vessels have machine-operated rods.

- **Long-line (pelagic or bottom)**—A boat extends a long main line either suspended at a chosen depth from floats or laid along the bottom. Shorter lines, called snoods, are attached to the main line at intervals; each snood ends in a baited hook. Each long-line can be miles long and may bear thousands of baited hooks.

- **Pot, creels, or traps**—This method is used for bottom-dwelling, continental-shelf species, such as crabs, lobsters, and octopus. Fishers place baited, specially designed traps on the bottom. The fish and mollusks enter the trap but cannot get out. The fishers retrieve the traps days later and harvest the fish and mollusks.

- **Dredging**—This method is used for shellfish, such as scallops, oysters, and clams. The fishers drop a special metal basket into the bottom mud. They drag the basket through the mud and lift it out of the water to retrieve shellfish. This is most often done in shallow waters such as bays and estuaries.

- **Dive catching**—Divers (snorkel, scuba, or hardhat) go into the water and individually catch the fish of interest. This is useful for coral reef fishes, such as grouper and snapper, and mollusks/shellfish, such as lobsters, scallops, abalone, and giant clams.

- **Harpooning or spearfishing**—A fisher or diver catches the target fish with a barbed spear. Harpooning is used in commercial whaling and some fishing (tuna, swordfish). Spearfishing is mainly used in dive catching and sport fishing.

- **Dynamite/cyanide**—While not practiced on a large scale, fishers in some places use explosives or poisons to stun the fish. These techniques result in enormous damage to marine habitats and are typically illegal, but still continue.

An important consideration in commercial fishing is selectivity, or how many of the fish caught in nets are the desired fish versus other types that wander into them. Some fishing methods are highly selective (harpooning, pole and line) by taking only the target fish, while others are nonselective (most netting, trawling, and long-line), resulting in a high death rate for non-target species, or "bycatch." Another consideration is that many net fishing techniques also ensnare marine mammals, such as dolphins, seals, and whales, and other endangered species such as marine turtles.

Commercial fishing has become highly technical and efficient; fleets are able to make large catches on single trips. This helps keep costs down, as diesel fuel is a major expense for commercial vessels. However, the ability to make large catches often leads to **overfishing**, that is, taking so many fish that too few are left to maintain the population. Overfishing can lead to extinction of marine fish species and diminish once prosperous fishing areas (e.g., Grand Banks off the coasts of New England and Newfoundland), and is the single greatest contributor to the decline of global fisheries. In efforts to reduce overfishing, many countries have entered into treaties that define how far domestic and foreign fishing fleets can fish off the coast of a given country (5 to 200 miles). The most important of these treaties is the United Nations Convention on the Law of the Sea, or Law of the Sea for short. Coastal states and countries may also set catch limits on fishing fleets operating within their jurisdiction.

One possible alternative to overfishing is **aquaculture** or **fish farming**. Fish farming involves raising populations of fish in ponds or fenced areas of estuaries and bays. The fish are fed, the population grows, and the population gets harvested in a sustainable manner. Aquaculture provides about one-third of the seafood we eat, and this portion will likely increase as human population grows. In addition to fish, shellfish such as oysters, clams, and mussels can be farmed and can provide benefits in cleaning local waters. However, aquaculture does have the following disadvantages:

- **May depend upon wild fish**—Depending upon the type of cultured fish, aquaculture requires wild fish for eggs and processed fish for food. Carnivores such as salmon and tuna require processed fish food. This places demands on commercial fishing. Vegetarian fish such as tilapia are good for aquaculture.

- **Crowding**—Raising thousands of fish in small enclosed areas creates problems such as the following:

  — *Wastes*—The fish generate huge amounts of fecal material that pollute the waters. The government regulates how these wastewaters must be treated.

  — *Disease*—Crowding can hasten the spread of diseases among the farmed population and into wild fish that come into contact with the same water. Treatment with antibiotics can select for antibiotic-resistant bacteria in the water.

- **May invade native habitats**—Farmed fish may escape and compete with, interbreed with, or take over habitats from native wild fish.

Whether cultivated or harvested from the wild, fish represent a renewable resource that, if properly managed, could help meet the food requirements of a growing population.

PUBLIC AND FEDERAL LANDS

The federal government owns and has set aside about one-third of the land in the United States for public use, public recreation, and wildlife conservation. (See Figure 6.10.) Most of these public lands are in the western United States and Alaska. The public lands include the following:

- **National Park System**—National parks, historic sites/trails, preserves, lakeshores, rivers, seashores, and recreation areas. The National Park Service manages these areas.

- **National Wildlife Refuge System**—Habitats and havens for wildlife from Alaska to Florida. The U.S. Fish and Wildlife Service manages these lands.

- **National Forest System**—The U.S. Forest Service manages national forests and grasslands for mixed—commercial and recreational—uses.

- **National Resource Lands**—The Bureau of Land Management (BLM) operates all federal lands that are not designated as forest, parks, or wildlife refuges. These lands are areas that no one wanted during the settlement of the west and Alaska. They also have commercial and recreational uses.

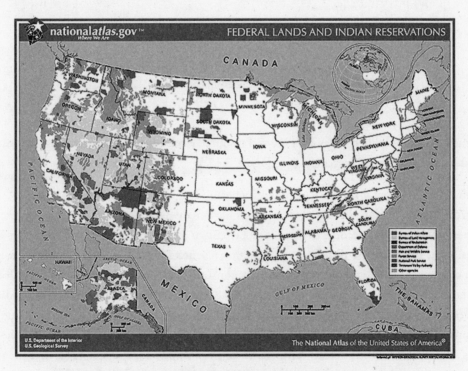

Figure 6.10 Map of U.S. Federal Lands and Indian Reservations
Source: Courtesy USDI/USGS.

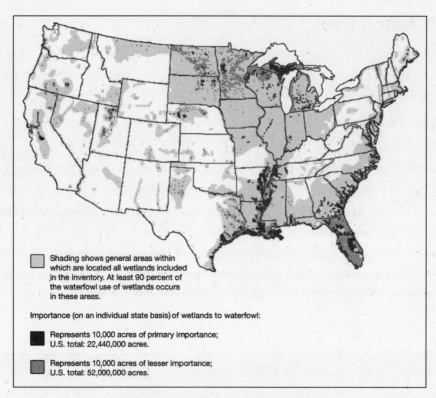

Shading shows general areas within which are located all wetlands included in the inventory. At least 90 percent of the waterfowl use of wetlands occurs in these areas.

Importance (on an individual state basis) of wetlands to waterfowl:

Represents 10,000 acres of primary importance; U.S. total: 22,440,000 acres.

Represents 10,000 acres of lesser importance; U.S. total: 52,000,000 acres.

Figure 6.11 Map of U.S. Wetlands
Source: Courtesy of USGS.

Within these parks, wildlife refuges, forests, and resource lands lie most of America's wilderness lands and many of America's wetlands. (See Figure 6.11.)

The public lands are used for diverse purposes including logging, mining, grazing, wildlife preservation, and recreation. Because federal lands have many uses, conflicts arise among the users. These conflicts have become divided into two major philosophies: wise use and strict conservationist. The **wise use philosophy** holds that resources on public lands should be used for economic purposes. This philosophy is widely consumptive for commercial uses (trees for logging, land for mining, and grasslands for grazing). Commercial users often lease the lands at minimal costs and obtain resources from them at little expense (i.e., well below market value). They subsequently make and sell products from these resources at great profits. In effect, the taxpayers subsidize these commercial ventures. In addition, these commercial users do little to help reclaim, conserve, or manage the lands, which ultimately leads to degradation of the ecosystems and habitats on federal lands. In contrast, the **strict conservationist philosophy** holds that the federal lands should be protected from commercial ventures. This philosophy would ban all logging, mining, and grazing, and would promote more sustainable uses of these lands. Obviously, some compromises must be established between these diametrically opposed philosophies, and sustainable management methods should be employed by the government agencies responsible for managing public lands.

An increasingly popular perspective on land use is the idea of **non-use,** where the preservation of the land leads to other non-exploitative uses, such as whale watching or birding. This concept applies not only to public lands but also to coastlines, oceans, and other common zones. A related concept is called **existence value,** where people derive satisfaction from the mere fact that certain habitats or wild animals exist.

All of these factors may be taken into account in a Cost-Benefit Analysis (CBA), an exercise that natural resource managers or developers perform when deciding whether to go ahead with an industrial project. Similarly, conservationists may perform a CBA to determine if a conservation project is the best use of resources. The projected benefits of a project, such as economic development and infrastructure improvements, are weighed against the potential costs, such as environmental degradation. A major problem is that it can be difficult to quantify many of the costs or benefits, such as the dollar value of lost ecosystem functions or of existence value. An entire field, Natural Resource Accounting (NRA), has emerged to handle these difficult calculations, with the ultimate goal of helping people make informed decisions about land use.

The major problems facing federal lands include the following:

- **Overuse**—Both commercial ventures and recreational users place enormous stresses on the habitats and ecosystems on federal lands, stresses that could be remediated by issuing fewer and more costly permits and leases. This would reduce the number of commercial and recreational users.

- **Commercial exploitation**—As long as it is cheap to harvest resources from federal lands, commercial exploitation/overuse will continue. But if the price of harvesting resources on federal lands increases, fewer enterprises will use them. Also, recycling will reduce the waste of paper and metal products and reduce the need for extracting timber and minerals from federal lands.

- **Pollution**—Air and water pollution from neighboring areas degrade ecosystems on federal land. Efforts must be made to minimize pollution and clean up polluted areas, both on and off federal lands.

- **Habitat destruction and threats to wildlife**—Pollution, grazing, mining, and timber harvesting destroy habitats. Increased tourism disturbs wildlife on federal lands. Decreased use of the land will help preserve habitats for wildlife.

- **Poor management**—Many federal agencies responsible for managing public lands are insufficiently funded and staffed. They have difficulties enforcing laws on federal lands and properly employing sustainable management techniques. But they could generate revenues from public lands—increasing the costs of using federal lands (permits, leases) and allowing the government to profit from the sale of resources. They could also implement sustainable resource management and conservation procedures. Both users and government agencies could cooperate on land reclamation and reforestation.

ENDANGERED SPECIES ACT

In 1973, Congress passed a landmark piece of legislation called the Endangered Species Act. It authorizes government agencies (Fish and Wildlife Service, National Marine Fisheries Service) to identify and list endangered and threatened species, both terrestrial and aquatic. By law, these species cannot be killed, collected, hunted, or injured within the United States and its territories. The decision to add or remove a species from the list must be based solely on biology, not economics. Under the act, federal agencies cannot implement, fund, or authorize projects that would harm an endangered or threatened species or destroy or change its habitat (land, air, water). Once listed, the Fish and Wildlife Service or the National Marine Fisheries Service must come up with a plan to help the endangered species recover. However, lack of funds means that recovery plans are not always made and implemented. Furthermore, there have been many attempts by legislators to weaken the Endangered Species Act; under political influence, legislators have tried unsuccessfully to amend the law so that economic factors must be considered in listing a species as endangered.

HOW POPULATIONS CHANGE THE LAND

People tend to live together in cities. As cities grow, they consume resources and land, while producing vast amounts of wastes. In addition, city growth destroys wildlife habitats. Cities and their growth must be carefully managed to sustain the land and its ecosystems.

URBAN LAND DEVELOPMENT

As the human population increases in industrialized countries, MDCs, and LDCs, people tend to migrate toward cities and urban areas where economic opportunities exist, such as employment, trade, and education. These migrations increase the sizes of cities and even cause cities to merge into huge **megalopolises**.

In the United States, seven huge urban areas exist across the country (see Figure 6.12):

1. Boston to Washington corridor—Boston, New York, Philadelphia, Baltimore, and Washington, DC

2. Pittsburgh to Chicago corridor—Pittsburgh, Akron, Cleveland, Toledo, Detroit, Indianapolis, and Chicago

3. Los Angeles

4. San Francisco Bay Area

5. Atlanta

6. Dallas, Fort Worth, Houston

7. Greater Miami and Florida coast

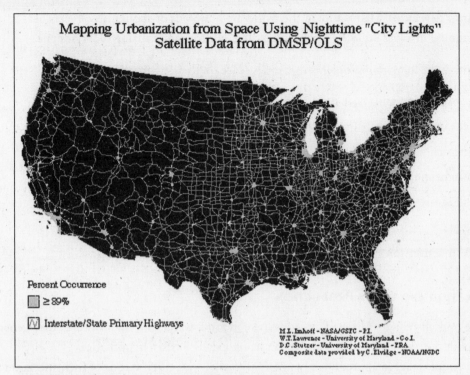

Figure 6.12 Urbanization in the United States.
Satellite image of city lights at night overlaid with a map of the U.S.
highway system.

Source: Courtesy of USGS, available at http://biology.usgs.gov/luhna/images/fig3_1.gif.

Let's look at how cities grow and develop. Following are the three major models:

1. **Concentric circle**—The business district occupies a core region in the center; there is more business than residents here. The city grows outward in rings of suburbs where residents live. Most people in the suburbs work in the central business district, while some businesses (mostly services) exist in the suburbs to support the residents. New York City is an example of a concentric circle city.

2. **Sector city**—The business district is in the center, but pie-shaped wedges of various residential (high-rent, low-rent, suburb) and industrial districts extend outward along major routes of transportation. San Francisco is an example of a sector city.

3. **Multiple nuclei city**—There are several independent centers of business and residential districts that merge to form one city, but do not show any particular pattern. Los Angeles is an example of a multiple nuclei city.

Cities tend to grow outward from the central area. If they cannot grow outward (because of geographical barriers such as mountains), then they will grow upward. There will be many highrise buildings for residential apartments as well as businesses. Hong Kong and Tokyo are examples of this type of urban development. One aspect of urban development is that rarely do people get to plan entirely new cities. Many cities in the world are hundreds or thousands of years old. So, most urban planning involves building upon, improving, and/or modernizing older designs.

Let's examine the following problems that cities place on their environments:

- Consumption and waste production
- Lack of vegetation
- Microclimates
- Water runoff
- Solid waste and pollution
- Noise
- Land conversion

CONSUMPTION AND WASTE PRODUCTION

Cities are not self-reliant entities. They must take in food, water, and fuel. In turn, they generate sewage, pollution, and solid waste. For example, on a daily basis, a hypothetical U.S. city of 1 million people takes in approximately 625,000 tons of water, 9,500 tons of fuel, and 200 tons of food. Each day, this hypothetical city generates 500,000 tons of sewage, 950 tons of air pollution, and 9,500 tons of solid waste. The inputs come from the surrounding environment and the outputs go to the environment. Thus, cities place great stresses on their environments. Cities can reduce stresses to the environment and benefit if individuals practice water conservation methods, reduce individual vehicle use, use public/mass transportation (when available), and recycle appropriate solid wastes (glass, aluminum, paper).

LACK OF VEGETATION

Most cities have few trees, forests, or other vegetation. Trees are important for giving off oxygen, cooling the air, absorbing pollutants, and providing habitats for wildlife. Vegetation can help control water runoff and soil erosion. Vegetation can also provide food, which otherwise has to be imported into the city. Some solutions to these problems include the following:

- **Planting trees on streets or empty lots**—Trees are aesthetically pleasing and provide valuable cooling (shade, evaporative cooling).
- **Planting gardens on individual lots or rooftops**—Small gardens can provide vegetables for individuals or be sold in "farmer's markets."
- **Establishing more parkland within the city**

MICROCLIMATES

City pavements absorb heat from the Sun. Tall buildings disrupt air currents. Cars, factories, and lights from buildings generate heat. Water runs off pavements rapidly and gets removed by sewage systems. These factors combine to give cities unique microclimates; cities tend to be warmer, cloudier, rainier, and foggier than surrounding areas (suburbs, rural). Cities tend to form a **heat island** that traps dust and pollutants (smog). The effects of heat islands can be minimized by planting trees, passing laws to have streets and buildings made from reflective materials, improving mass transit, and having high efficiency energy standards for buildings, industries, and cars.

WATER RUNOFF

As mentioned above, cities place great demands on local water supplies and must build systems to deliver water to them (e.g., dams, reservoirs, aqueducts, canals). A city can deprive surrounding rural areas of needed surface waters (some of which also help to replenish groundwater). In some cases, these demands can be placed on aquifers even farther away. For example, Los Angeles and Southern California take much of their water from the Colorado River through an elaborate system of aqueducts; this water not only meets the drinking needs of Southern Californians, but it also satisfies the irrigation needs of Southern California farmlands.

The city pavements form barriers that prevent the underlying ground from receiving moisture, which in turn degrades the quality of the underlying soil and depletes groundwater. City pavements contribute to water runoff. The water in this runoff carries pollutants from the pavements, such as oil and toxic chemicals, that eventually make their way through the storm sewage system into streams and rivers.

SOLID WASTE AND POLLUTION

As previously mentioned, a hypothetical U.S. city of 1 million people generates 9,500 tons of solid waste daily. The single largest component of trash is paper and paper products, followed by plastic, but some surprising items, such as cleaning products, make up a substantial proportion of household waste. Some of this trash gets burned, but most of it ends up in landfills. Most city landfills are located outside of the city in surrounding areas. Although most municipal solid waste landfills are designed to isolate the trash from the environment for long periods of time, the success depends on how well the landfill is constructed and operated. Water must be prevented from passing through the landfill with storm drainage systems, and water that does go through the landfill (leachate) must be collected and treated before being discharged into local rivers and streams. Because cities generate such large amounts of trash daily, existing landfills eventually fill up and new ones must be built in surrounding areas. For example, the Fresh Kills Municipal Landfill that served New York City from 1948 to March 2001 was built on 2,200 acres of

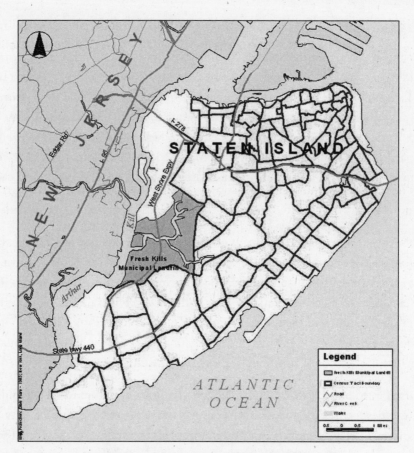

Figure 6.13 Fresh Kills Municipal Landfill
Source: Photo courtesy of CDC.

Staten Island and was once the largest landfill in the world. (See Figure 6.13.) It was reopened after 9/11 to receive debris from Ground Zero. New ideas for the area have included turning it into reclaimed wetlands, recreational facilities, and landscaped public parkland.

To reduce trash and help landfills last longer, city residents (and others) can recycle items, such as plastic, glass, aluminum, and paper, and compost kitchen wastes, such as vegetables and eggshells. Recycling is especially important since many common waste products take a very long time to decompose. Aluminum cans, for instance, take 200–500 years to break down. Paper often takes only a few weeks to decompose, but under certain conditions it can persist for decades. Compost can be used to fertilize municipal trees and gardens or individual gardens.

Sewage treatment plants must be able to handle and treat the wastewater generated by cities. Otherwise, waterborne diseases such as dysentery, cholera, and typhoid will spread. In many older cities around the world, especially in LDCs, population growth in cities can exceed the capacity of their waste treatment plants.

As mentioned, both air and water pollution are major problems in cities. Runoff contributes to water pollution, and automobiles and industrialization contribute to air pollution. Regulations must be implemented to help reduce the effects of pollution. Use of mass transit rather than individual automobiles will help as well.

NOISE

City dwellers are constantly exposed to high levels of noise from cars, trucks, buses, construction, sirens, and the everyday hustle of city life. High levels of noise can lead to permanent hearing loss. Shielding noisy devices (walls along major highways will reduce the noise from passing traffic), moving noise sources away from people, and using new antinoise technologies (new construction equipment) can reduce noise levels significantly.

LAND CONVERSION

As cities grow outward, rural land gets converted into urban/suburban areas. This reduces rural forests or agricultural land and, in turn, brings all of the urban problems to new areas. Development of the surrounding areas can increase property values and lead to higher taxes for the rural residents. Furthermore, cheap gasoline prices, abundant land, and good highway systems can lead to a spread-out or dispersed type of development (low population density) called **urban sprawl**. In urban sprawl, people must use cars to go practically everywhere (shopping malls, grocery stores, work, etc.). The increased car use contributes to traffic problems, noise, and pollution. Urban sprawl converts more rural land than necessary for city growth. Good urban planning, zoning, and regulation of development are necessary to control urban sprawl and city growth.

Ecological land-use planning takes the following geological, ecological, economical, sociological, and health factors into account when planning growth:

- **Survey geology, ecology, sociology, and health of an area**—Areas vital for supplying drinking water, minimizing habitat destruction, and preventing soil erosion are identified and protected.

- **Ask about the goals of development and their importance**—Are economic development and population growth to be encouraged or discouraged? Is the goal to prevent soil erosion?

- **Develop composite maps**—Combine data to develop an overlay of geologic, ecologic, and socioeconomic maps of an area.

- **Develop and implement a master plan**—Use data from the composite maps to make decisions about which areas should be developed.

Ecological land-use planning is great in theory, but it is not usually implemented because it is costly. There are also political and economic pressures for short-term gains, and land-use planning requires mutual cooperation between municipalities.

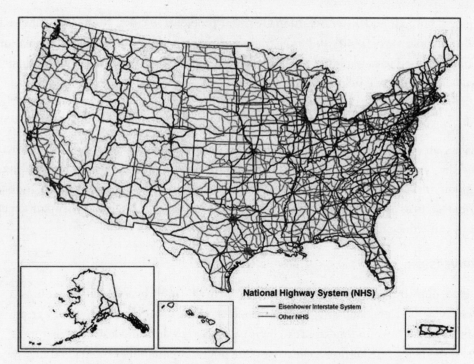

Figure 6.14 Map of the U.S. National Highway System
Source: Courtesy of FHA/USDOT.

TRANSPORTATION INFRASTRUCTURE

As you can see in Figure 6.14, the federal highway system has increased access to many areas of the United States.

Many roads traverse diverse ecosystems, and the construction of these roads has had the following impact upon them:

- **Construction degrades the environment**—Strips of forest may be cleared to make way for a road. Bridges need to be built across valleys or bodies of water. In some areas, tunnels must be dug through mountains. Deforestation can lead to disruption of habitats and soil erosion, and construction equipment can cause pollution.

- **Urbanization**—When you access new areas with roads, urbanization (urban sprawl) tends to follow. If you look at Figure 6.12, you can see how urban centers are closely linked with roadways.

- **Pollution**—Increased road access leads to more traffic.

Similar problems arise when new waterways such as canals and channels are built. New waterways create new commerce areas that lead to urbanization. Furthermore, channels and canals may divert the usual flow of water from a river and impact the wetlands downstream. For example, shipping channels and flood control dams created along the Mississippi River through Louisiana by the U.S. Army Corps of Engineers alter water flow and sediment flow patterns. The U.S. Army Corps of Engineers has engaged a number of projects for flood control, navigation, sediment diversion, and wetland reconstructions along the Mississippi River, the Mississippi delta, and adjacent rivers in Louisiana.

HOW POPULATIONS CAN SAVE THE LAND

As you can see, the use of the land by humans for mining, logging, grazing, urban projects, and farming can have damaging consequences. How can land be saved? Following are three basic approaches:

1. Don't use the land at all—preservation

2. Use the land, but fix it afterward—remediation, restoration

3. Use the land wisely and ecologically—sustainable land-use strategies

Let's look at these approaches.

PRESERVATION

The idea of preservation is to prevent the land from being used. The following methods can be used for preserving land:

- Purchasing

- Legal agreements

- Donations

- Debt for nature swaps

One method of prevention is to purchase the land. As noted in the Public Lands section, the federal government owns almost one-third of the land in the United States and uses it for various purposes (conservation, mining, forestry, recreation). Besides purchasing land directly, governments (federal, state, and local) may preserve land by passing zoning laws that regulate how land is to be used or preserved. Besides governments, individuals and organizations (e.g., Audubon Society, Nature Conservancy) may preserve lands by purchasing them **(land acquisition)** and designating them for conservation.

Another method of preserving land is to place legal restrictions on how it can be used. One such method is a conservation easement. In a **conservation easement**, the landowner agrees to protect some portion or all of his land from residential, commercial, or industrial development in exchange for money or tax breaks (income, property, estate, etc.). Governments and private organizations can enter into conservation easements with private landowners. Basically, the government or organization owns the developmental rights to the land under easement, but the landowner maintains full title and ownership of the land. The conservation easement applies to the property regardless of whether the property gets sold or passed on to heirs. Conservation easements are commonly used to protect and preserve agricultural lands, historical sites, scenic places, and lands rich in natural resources.

Some private landowners also donate their land to conservation organizations for protection. The owners may receive tax breaks or avoid taxes (e.g., estate taxes) by donating the land. The conservation organization then owns and manages the land.

In some international cases, private organizations may help LDCs with sensitive ecological areas eliminate foreign debts by conserving those areas. For example, The Nature Conservancy has a **Debt for Nature Swap Program** in which this organization works with the U.S. Treasury Department and foreign governments. They raise money and work with the U.S. government to provide money that will be used to forgive a foreign country's debt in exchange for purchasing lands in the foreign country to be managed by a conservation group(s). The Nature Conservancy has brokered Debt for Nature Swap agreements for preservation of tropical rain forests between the United States and the countries of Belize, Colombia, Panama, and Peru.

REMEDIATION AND RESTORATION

Remediation and restoration are two processes involved in what may be generally known as **disturbed land restoration** or **ecological restoration**. Basically, disturbed land restoration is the process by which land that has been altered by development, road construction, agriculture, mining, or grazing is returned to pre-disturbance conditions. These human activities can disrupt ecosystems (e.g., destroy soil, alter water pathways, introduce exotic species, erode soil, and deposit sediment) and leave long-lasting pollutants in the water and soil of the environment. To restore an area requires understanding of the biological and physical components of that particular area. What type of soil was present before the disturbance? What was the pattern of water flow through the area? What plant and animal species lived in the area? Once you recognize and understand the physical and biological components that were disturbed, you can begin to correct them. Because the process of ecological restoration may take decades, re-creating the entire undisturbed system is impractical. Rather, you should eliminate the unnatural components and ensure that the natural recovery processes are allowed or induced to happen. Eliminating the unnatural components could be as simple as removing a roadbed or as complicated as decontaminating the soil (remediation). Once the unnatural components are removed, then you devise a plan to allow natural resources to recover (replacing soil, planting vegetation, restoring stream flows, introducing animals).

Remediation is the removal of pollutants from the disturbed land, including sediments in waterways. There are numerous methods and the type of method used will depend upon the type and nature of the contaminant, the extent of the contamination, the potential expense of remediation, and the available funding for remediation. Methods of remediation and restoration include chemical, biological, and physical techniques. For example, an area may be revegetated. New species may be introduced, or undesirable ones removed. An area may be graded to return the natural topography. Irrigation ditches may be refilled and streams diverted to restore the natural pattern of water flow. New sediments or soil may be added to replace soil that was removed by mining or remediation. Chemical nutrients may be added to the soil to restore depleted ones. Remediation falls into the following categories:

- **Containment**—Isolating the contaminants from the rest of the ecosystem through use of the following:

  — *Landfill caps or brownfield caps*—plastic or concrete barriers used to cover the affected area and divert water away from it

  — *Stabilization or solidification*—cement, concrete, or chemicals added to the area to bind and immobilize the contaminants

- **Destruction/transformation**—Destroying the contaminant or changing it into some less harmful substance by doing the following:

  — *Bioremediation or biodegradation*—bacteria, fungi, or microbes introduced into the area to change the contaminant from a toxic to a nontoxic form

  — *Chemical oxidation-reduction*—injecting chemicals that react with the contaminant into the area; the chemical reactions change the contaminant to a nontoxic form.

  — *Bioventing*—injecting oxygen into the area to improve the activities of aerobic microorganisms that can change the contaminants

  — *Dehalogenation*—injecting chemicals to remove halogens such as chlorine and fluorine

  — *Incineration*—burning contaminated soil to destroy the contaminant; usually done after the soil has been removed

- **Removal**—Taking the contaminant from the area through the following means:

  — *Off-site disposal*—removing and storing the contaminated soils and sediments in a sanitary landfill

  — *Thermal desorption*—injecting steam into the ground to extract pollutants with low boiling points (volatile organic compounds such as petroleum)

  — *Soil washing*—removing soil and washing with water to remove oils and heavy metals

  — *Phytoremediaton*—introducing plants to the contaminated soil. As the plants grow, they take up the contaminant (e.g., heavy metal) from the soil and incorporate it into their tissues. The plants are then harvested, burned, and the pollutant recovered. In some instances, phytoremediation may also be used as a bioremediation method.

— *Filtration*—contaminated water flows through a porous substance (e.g., sand, clay), removing solid particles from the liquid

— *Ion exchange, carbon adsorption, reverse osmosis*—contaminated water flows through columns with chemicals that bind or remove the substance from the water

— *Fluid/vapor extraction*—liquid and gases vacuumed from the area carry volatile organic compounds with them

Once the unnatural components are removed, restoration processes can begin. Again, restoration may take years or decades of active management before the site returns to its natural or reference state.

SUPERFUND AND TOXIC CLEANUPS

In the 1970s, several areas of the United States were discovered to be highly polluted with hazardous wastes, particularly the Love Canal area of New York. In 1980, Congress passed the Comprehensive Environmental Response, Compensation, and Liability Act that authorized the federal government to clean up hazardous waste sites and respond to accidental or disastrous releases of hazardous wastes into the environment. The law sets regulations and prohibitions regarding closed or abandoned hazardous waste sites, provides for liability of people and companies who cause hazardous waste discharges or dumping, and establishes a multi-billion-dollar pool of money dedicated for hazardous waste cleanup (Superfund) when no responsible party can be identified. The Environmental Protection Agency manages the Superfund. The law also provides for short-term removals and long-term remediations. Several Superfund sites were created across the country to dispose of these wastes by burying them in sanitary landfills. While the Superfund was plagued with initial technical problems (leakage from Superfund landfills) and insufficient funding, it has had several successes in its 26-year history. The Superfund has been reauthorized several times since its initiation and continues to be the nation's major response mechanism for handling hazardous waste spills.

SUSTAINABLE LAND-USE STRATEGIES

It may seem from the discussion so far that there is a constant battle between environmentalists and economic interests over using land, whether for activities such as farming, timber, mining, or grazing. Historically, this has been true and continues today. You can see it in the discussions about the oil in the Arctic National Wildlife Refuge (ANWR) that was debated in Congress during the administration of President George W. Bush. Obviously, we as citizens cannot ignore the human activities and economic interests of our country by not using the resources that it offers, yet we also cannot ignore our environment by destroying ecosystems to extract those resources. Some compromise must be reached. This compromise is **sustainable land-use strategies**, which basically means using ecological principles to guide land-use decisions. The Ecological Society of America (ESA) published a paper entitled "Ecological Principles and Guidelines for Managing the

Use of Land" in *Ecological Applications,* Volume 10 Number 3 (June 2000). In this paper, the ESA outlined the following ecological principles that must be recognized in land-use decisions:

- **Time principle**—Various ecological processes function on different time scales from short (hours, days) to long (years, decades, centuries). In addition, because ecosystems change over time (e.g., seasonally), their structures and functions are a reflection of their history. Furthermore, the effects of human activities on ecosystems may not be immediately noticeable, may last for long periods of time, and may be difficult to predict.

- **Species principle**—Interactions of species within an ecosystem have important broad effects on that ecosystem. Indicator species, keystone species, and trophic levels are important in the functioning of the ecosystem. Some species, like beavers, alter the ecosystem through their activities. Human activities can alter the ecosystems in ways that change the composition and diversity of species.

- **Place principle**—Local climate, water, soil, geology, and biologic interactions strongly affect both the functioning of the ecosystem and the abundance and distribution of species in any one place. For example, rainfall can limit which plant species can thrive in an area. Only those species adapted to a specific place can and should thrive there. Altering the landscape to make way for other non-native or non-adapted species (e.g., non-native crops supplied by irrigation) may be too costly ecologically and economically in the long term.

- **Disturbance principle**—The characteristics of the disturbance (type, intensity, duration) shape the characteristics of the ecosystem (populations, communities, the ecosystem itself). Disturbances may be natural, such as wildfires or floods, or human-made, such as clear-cutting, road construction, dam construction, or mining. Some natural disturbances may be important in maintaining the ecosystem (wildfires, floods), but may be costly to human activities. Human activities such as flood control dams and wildfire breaks may alter these natural disturbances, thereby altering the ecosystem.

- **Landscape principle**—Land cover (size, shape, spatial relationships) influences the populations, communities, and ecosystems. Disrupting the spatial distribution of land cover like the clear-cutting of forests creates gaps and fragments in the ecosystems across which many species cannot interact.

With these principles in mind, the ESA has suggested the following eight guidelines for ecological land use:

1. **Study the impact that local land use decisions will have on the region**—What is the regional context (soil types, land cover, water movements, habitats, species)? What disturbances have happened in the past? How are the lands adjoining the region used? Based on the regional context, what uses could be made for the land that might satisfy ecological constraints as well as socioeconomic and political concerns?

2. **Make long-term plans as well as plans for unexpected events**—What physical, biological, and economic constraints will the proposed land-use decisions have? What might be the effects of these decisions over long periods of time? What might be the effects of land-use decisions made on adjoining lands on the region in question? How will the land-use decision be affected by events such as floods, hurricanes, or droughts?

3. **Preserve rare landscape elements and species**—What features of the landscape, soil, water, or biological communities are rare? Where do they occur in the region? Sometimes, rare communities have broad effects on the ecosystem as a whole. Can these areas with rare species or features be preserved, while still making use of the land (e.g., wildlife corridor)?

4. **Avoid land uses that deplete natural resources over a broad area**—Will the land use cause soil erosion over a large area? Will the land use destroy wetlands and their associated species? Will irrigation place too high a demand on natural water resources for this area? Perhaps planting crops that grow in a drier climate would be more appropriate. Are the proposed land uses appropriate for this particular setting at this particular time?

5. **Keep large areas or connected areas with critical habitats**—One large habitat area may be better than many smaller fragmented ones.

6. **Minimize non-native species**—Non-native species usually harm native species and the functioning of the ecosystem. They usually do not have the normal checks on their growth, such as predation, that native species do. Land-use planners should encourage the growth of native species that are best adapted to the environment.

7. **Avoid/compensate for the negative effects of development on the ecology**—Think about placing land-use structures that will have the least ecological impact, if possible. If not, can you create areas within your structures for wildlife habitats (e.g., parks, gardens, golf courses)?

8. **Make land-use management practices that are compatible with the ecological potential of the area**—Is the region more appropriate for forestry or grazing than for farmland? If you can match the land use with the potential of the area, it will be more cost-effective and ecologically sound.

Sustainable land-use strategies require a great deal of study and planning. It is important to understand the ecology of the area before decisions are made and to make decisions that are consistent and minimally disruptive of the ecology. Environmental scientists, politicians, business leaders, and community members should be involved in the decisions. Always remember that human activities have long-lasting effects on the environment.

ENVIRONMENTAL LEGISLATION

Laws and regulations are an important component of preserving and protecting the land. Laws grant public agencies authority over previously ungoverned resources and commerce. Laws also allow actions on behalf of the common good that might go against individual interests. Laws take the responsibility for making unpopular decisions, such as restricting hunting, and place it with objective authorities rather than leaving management of communal natural resources up to individuals. (For more on this, see the discussion of the "Tragedy of the Commons" in Chapter 9.)

Consider the example of the wildlife trade in the United States. In the late 19th century, the trade in wildlife products, particularly feathers, was extremely profitable. With bird populations dwindling, hunters rushed to kill the birds before their competitors, to try to reap the remaining profits. The Lacey Act, passed in 1900, created oversight over this trade, whereas previously each individual acted as he or she saw fit. Even when an action is clearly detrimental to the community, such as shooting every last bird, without laws and regulations the community has no way to restrict the action. The laws needed to protect community resources have been updated through the years to keep track with evolving attitudes, technology, and scientific understanding. Table 6.2 lists several of the most important U.S. environmental laws and their intended goals.

Table 6.2 Selected Major U.S. Environmental Laws

Name	Year Passed	Major Provisions and Notable Features
Rivers and Harbors Act	1899	Made it illegal to pollute the water and dredge a water body without a permit; notable as the first environmental law in the United States
Lacey Act	1900	Banned commerce in illegally harvested animals or plants
Antiquities Act	1906	First law to establish penalties for disturbance of archaeological sites or relics on public lands; gave the president the right to designate landmarks and other areas of "historical or scientific interest"
Migratory Bird Treaty Act	1918	International agreement that outlawed, in part, hunting of or trade of designated migratory birds
Clean Air Act	1963; revised repeatedly, notably in 1970 and 1990	Set air pollution standards, which the EPA was later required to enforce; established the right of private citizens to sue under this act; later amendments added provisions for acid rain, ozone depletion, and trading emissions credits
National Environmental Policy Act	1969	Required all federal agencies to provide Environmental Impact statements for their activities or regulatory actions; established President's Council on Environmental Quality; led to creation of the EPA by executive order. All its provisions also apply to cultural heritage

Clean Water Act	1972	Established standards of water quality; established permit requirements for point sources of water pollution; required and provided funding for municipal sewage treatment; provided for research on nonpoint source pollution; required EPA and states to compile a national inventory of water bodies and their condition
Noise Control Act	1972	First law to regulate and establish standards for noise pollution
Endangered Species Act	1973, but built on earlier laws, including the Lacey Act; amended repeatedly	Gave the Department of the Interior the authority to list endangered species of plants and animals, which may not be killed or harassed in any way, and whose habitat must be protected; directed federal agencies to preserve habitat for endangered species on lands they manage and avoid any action that will harm listed species; established rules to ban or prohibit interstate or international trade in endangered species (CITES treaty); required responsible federal agencies to establish recovery plans for endangered species; required all commercial wildlife trade to go through designated ports
Safe Drinking Water Act	1974	Required EPA to set standards for drinking water and groundwater, and provide oversight of the enforcement of these standards
Surface Mining Control and Reclamation Act	1977	Established environmental standards and permit requirements for surface mines; gave enforcement authority over mining operations to federal regulators; banned mining entirely in certain public lands; established Abandoned Mine Land Fund to pay for reclamation of abandoned mines
Comprehensive Environmental Response, Compensation, and Liability Act (CERCLA, aka Superfund)	1980	Authorized the EPA to identify parties responsible for contaminated areas and compel them to clean up the damage at their expense; established a trust fund, paid for by taxes on polluting industries, to clean up abandoned sites where the responsible party cannot be found
Oil Pollution Act	1990	Passed in response to the Exxon Valdez grounding in 1989; established that parties responsible for oil spills had civil and financial liability for damage and cost of clean-up, within set limits; required companies to have a plan in place to prevent or contain oil spills; established regional clean-up plans in the event of future spills; banned large tankers from Prince William Sound, Alaska

REVIEW QUESTIONS

1. Which of the following actions is consistent with sustainable agriculture?

   (A) Planting one crop only

   (B) Irrigating the fields

   (C) Applying heavy loads of fertilizers and pesticides

   (D) Planting alternating rows of multiple crops

   (E) Plowing the field under after harvesting

2. Which of the following is a nonrenewable resource?

   (A) Fish

   (B) Aluminum

   (C) Hardwood trees

   (D) Grasses

   (E) Agricultural crops

3. Which tree harvesting method is the worst ecologically?

   (A) Selective cutting

   (B) Shelterwood cutting

   (C) Clear-cutting

   (D) Seed-tree cutting

   (E) Slash and burn

4. What type of fishing uses a funnel-shaped net with doors to catch bottom-dwelling fish?

   (A) Trawling

   (B) Hook and line

   (C) Drift net

   (D) Purse-seine

   (E) Dredging

5. Legislating that products harvested on federal lands must return profits to the government would be the best away to address which of the following threats to public lands?

   (A) Overuse

   (B) Pollution

   (C) Habitat destruction

   (D) Poor management

   (E) Commercial exploitation

6. What is a conservation easement?

   (A) A purchase of land by a private organization for conservation purposes

   (B) A purchase of land by a government for conservation purposes

   (C) A legal agreement exchanging public debt for land for conservation purposes

   (D) A legal agreement where the landowner gives up development rights on his land in exchange for money or tax breaks

   (E) A legal agreement where a landowner agrees to let the government only develop the land in exchange for tax breaks

7. Scientists make a genetically engineered form of tobacco plant that can take up mercury and incorporate it into the leaves. They then plant these tobacco crops on a mercury-contaminated waste site, allow the plants to grow, and harvest them. Which of the following techniques fits this description?

   (A) Landfill or brownfield cap

   (B) Phytoremediation

   (C) Bioventing

   (D) Chemical oxidation/reduction

   (E) Biodegradation

8. Zoning is a good way to address which of the following problems with urban development?

(A) Water runoff

(B) Microclimate

(C) Land conversion

(D) Consumption/waste production

(E) Pollution

9. Which of the following would NOT be a principle for sustainable land use?

(A) Consider all aspects of land use (economic, ecologic, social, political).

(B) Take all the resources possible from a given area of land.

(C) Preserve biodiversity.

(D) Conserve resources through reduce, reuse, and recycle practices.

(E) Rehabilitate the land after use.

10. Which of the following is NOT affected by deforestation?

(A) Soil erosion

(B) Water cycle

(C) Global temperature

(D) Biodiversity

(E) Ozone layer hole

ANSWERS AND EXPLANATIONS

1. D

This question addresses your ability to recognize practices of sustainable agriculture versus conventional agriculture. Sustainable agricultural methods involve planting more than one crop, so choice (A) is incorrect; leaving the natural environment of the field, so choice (B), irrigation, is incorrect; and using integrated pest management techniques with little use of chemicals, so choice (C) is incorrect. That they are harvested at different times and the fields are left intact to minimize soil erosion means choice (E) is incorrect. Sustainable agriculture uses multiple crops, so (D) is correct.

2. B

A renewable resource can be replaced, while a nonrenewable resource cannot be replaced. When properly managed, fish, trees, grasses, and agricultural crops can be replaced and replenished, so choices (A) and (C) through (E) are incorrect. There is a finite supply of aluminum ore (bauxite) in the ground that cannot readily be replaced. Aluminum is a nonrenewable resource, so choice (B) is correct.

3. C

Of the tree harvesting methods, selective cutting, shelterwood cutting, and seed-tree cutting are all ecologically sound methods that attempt to spare trees of various ages and reproductive maturity to maintain biodiversity in the forest and to replenish the forest. Choices (A), (B), and (D) are therefore incorrect. Slash and burn is an agricultural method of clearing farmland, not a tree harvesting method, so choice (E) is incorrect. Clear-cutting eliminates all trees from an area or strips of adjacent areas. Clear-cutting does not attempt to preserve any area of the forest and is not an ecologically sound method, so choice (C) is correct.

4. A

Purse-seine and drift net fishing are for catching pelagic fish, so choices (C) and (D) are incorrect. Hook and line does not involve nets and is used for schools of pelagic fish like tuna, so choice (B) is incorrect. Dredging is used for bottom-dwelling shellfish and uses a metal scoop basket to pick up mud containing the shellfish, so choice (E) is incorrect. Trawling involves dragging a funnel-shaped net along the bottom to pick up demersal or bottom-dwelling fish, so choice (A) is correct.

5. E

Legislating that governments get profits from the sale of resources or products made from public lands would not directly address overuse, pollution, poor management, or habitat destruction on public lands, so choices (A) through (D) are incorrect. Many companies can rent or lease public lands cheaply and can make a large profit on the sale of resources or products from public lands. However, insisting that governments make a profit from resources on public lands and increasing the costs of doing business on public lands would directly address commercial exploitation, so choice (E) is correct.

6. D

A conservation easement is a legal agreement, not a purchase of land, so responses (A) and (B) are incorrect. A conservation easement does not involve swapping public debt for land for conservation, so choice (C) is incorrect. The land cannot be developed at all by anyone, so choice (E) is incorrect. In a conservation easement, a landowner gives up or sells the developmental rights to a government or private organization in exchange for tax breaks or money, so choice (D) is correct.

7. B

Landfill or brownfield caps isolate contaminated soils from the environment with barriers, so choice (A) is incorrect. Bioventing injects oxygen into the contaminated soil to facilitate degradation by microbes, so choice (C) is incorrect. Similarly, biodegradation injects microbes into the contaminated soil to facilitate the degradation of the contaminant, so choice (E) is incorrect. Chemical oxidation/reduction injects chemicals into the contaminated soil to transform the contaminant, so choice (D) is incorrect. Phytoremediation uses plants to take up the contaminant from contaminated soil, and subsequent harvesting of the plants removes the contaminant for disposal, so choice (B) is correct.

8. C

Zoning laws designate the purposes for which various areas of urban land can be used (residential, commercial, industrial). Water runoff and microclimates are physical urbanization effects on the environment and would not be altered by zoning laws, so choices (A) and (B) are incorrect. Consumption/waste production and pollution would not be affected by zoning laws, so choices (D) and (E) are incorrect. Land conversion would be directly affected by zoning laws, so choice (C) is correct.

9. B

Good practices of sustainable land-use strategies involve conserving resources, preserving biodiversity, planning/management with consideration for all interests (environmental, economic, political, social), and rehabilitating the land after use. So choices (A) and (C) through (E) are incorrect. Sustainable land-use strategies usually involve taking only what is necessary from the land, not everything, so choice (B) is correct.

10. E

Trees hold on to soil and play an important part in the water cycle because of evapo-transpiration, the movement of water through plants, so choices (A) and (B) are incorrect. From the atmosphere, trees take up significant amounts of carbon dioxide, which absorbs heat and influences global temperatures, so choice (C) is incorrect. Forests, especially tropical rain forests, provide habitats for many species of wildlife and are critical for preserving biodiversity, so choice (D) is incorrect. The ozone layer hole is caused by the interactions of human-made chlorofluorocarbons and oxygen in the upper atmosphere and has nothing to do with deforestation, so choice (E) is correct.

ANSWER TO FREE-RESPONSE QUESTION

For this question, you are expected to direct your response to a particular system but are required to draw on knowledge from several different areas to create a comprehensive answer. Key points for the answer to (A) include primary producers, hypoxia, dissolved oxygen, photosynthesis, respiration, and trophic levels. Begin with the primary producers and trace the path of oxygen through the ecosystem. A diagram may not be a bad idea here, but don't let the diagram stand alone. While composing this portion, you should be able to pinpoint trophic levels that would be affected greatly by hypoxia.

In (B) you should include aspects of the organisms' ecology outside of the oxygen cycle. Ideas to mention include organisms, primary producers, consumers, respiration, and energy flow in a community. If you are using an omnivore for your example, consider factors such as what organisms it eats, where it lives, and how it finds a mate.

(C) expands the negative impact on the oxygen cycle into the food web because sharks are top-predator carnivores. Mention key points like top predators, trophic levels, oxygen deprivation, and decreased food supply. Remember, sharks still need oxygen to breathe, but they also need nutrients to grow and for metabolism.

(A) Again, you can include a diagram, but your response should also mention the following points:

- Except for trace amounts of free oxygen that originate from the air-water interface, free oxygen exists as a result of photosynthesis by autotrophic organisms using energy from the Sun to fix carbon from atmospheric CO_2 while releasing H_2O and O_2.

- In the marine system, the primary producers are various forms of seaweed alga and pelagic photosynthe... plankton like diatoms. The oxygen these organisms generate is released into wa...r where some escapes to the atmosphere, depending on the atmospheric pressure ...ll the other members of the ecosystem—carnivorous and herbivorous fish, herbivo...us echinoderms, and detritivorous crustaceans, for example—utilize dissolved oxyg...n in the water table for metabolic respiration. Dissolved oxygen is also necessary for certa... chemical reactions involving the breakdown of organic tissue during decomposit... The processes of respiration and decomposition recycle the oxygen into CO_2.

- This cycle is upset in the Gulf of Mexi... because the nitrogen- and phosphorous-rich freshwater entering the marine system ...ports a surface-level algal bloom that blocks sunlight from reaching the natural prima...producers on the marine floor and in the water table. The biomass of the algae increa...beyond the system's ability to produce oxygen, so that the rate of O_2 production ca...eet the need for respiration and decomposition. As a result, the organisms nee...xygen and sunlight die until the algal bloom disappears. If enough of the organisms f...e natural system can survive, the system begins to recover until the return of the alg...om.

(B) The algal bloom affects the autotrophs most significantly by blocking sunlight. This not only prevents them from producing oxygen, but also prevents them from producing energy in the form of storage carbohydrates. All organisms in the ecosystem are connected somehow, and autotrophs are no different. As organisms in the ecosystem die, the resources they provide disappear. Autotrophs rely on corals for substrate anchoring points. They rely on predators to decrease herbivory. As oxygen levels decrease in the water because of lower photosynthetic rates, slow-growing corals die and predators leave the area, causing an increase in herbivory.

Herbivores are affected as the number of grazing spots decreases due to the death and decreased productivity of their algal food source. Herbivores also depend on their algal forest for protection from ocean currents and predators. As oxygen levels decrease in the water, herbivores with limited mobility, like sea urchins, suffer increased death rates, as they can't move quickly to more productive areas.

(C) Two factors may be contributing to a relocation of a top predator like a shark. First, the oxygen content of the water may be unsuitable for a shark population. Normal function is not possible when the oxygen supply is limited. Sharks that are mobile can migrate to an area that isn't affected or is less affected by the algal bloom.

The second factor is the food supply. As primary productivity decreases, it affects all levels of the ecosystem. Herbivore, omnivore, and primary carnivore species either die or migrate to a more suitable habitat. All of these trophic levels are part of the shark's nutrient reservoir. A shark would have to increase its hunting range to encounter more food items. A shark might migrate with its prey items if they move to a more suitable habitat. Finally, a shark might need to switch to a different food source if its naturally occurring sources of nutrients disappear. Unfortunately, human bathers may become an alternate food source for larger sharks that are normally not considered man-eaters.

CHAPTER 7:   ENERGY USE

IF YOU LEARN ONLY FIVE THINGS IN THIS CHAPTER . . .

1. Fossil fuels (oil, coal, and natural gas) currently supply roughly 86 percent of worldwide commercial energy. These fuels are easy and convenient to use, but they have environmental, political, and social costs.

2. To reduce or even eliminate dependence on fossil fuels or nuclear energy, these sources have to be used more efficiently and new technology for renewable sources needs to be relied upon.

3. Solar energy is a vast and significant resource. Solar energy is directly used in passive and active solar collectors. Solar energy is also indirectly responsible for other types of renewable resources such as wind power, hydropower, and tidal energy.

4. None of the renewable sources of energy discussed here are likely to completely replace fossil fuels. It is possible, however, to use a combination of alternative energy sources to make a significant difference in our energy needs.

5. Transportation is one of the biggest areas of energy use, but it is also the area that shows the most promise for conservation and new technology.

THE BASICS OF ENERGY

This chapter starts with a basic overview of what energy is and the different forms it can take on Earth in order to facilitate your understanding of energy consumption and the types of energy used by humans.

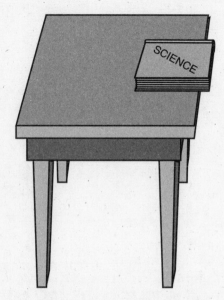

Figure 7.1 Potential Energy

Energy is the ability to do **work** or cause change. In many cases, energy is released in the form of **heat**. Even though we are all familiar with many different energy sources, such as light, heat, electricity, and chemicals, let's look more closely at the major types of energy.

Potential energy is stored energy that is available for use. The book on the table top seen in Figure 7.1 has potential energy just as water stored behind a dam has potential energy. The energy has not been released yet, but it *could* be released and used in another form.

Potential energy can be changed into **kinetic energy**. What would happen if that book were pushed off the table? The book would move and the energy would become the energy contained in moving objects. Kinetic energy is based on an object's mass and speed or velocity. Temperature is a measure of the kinetic energy created by the motion of atoms. As an object heats up, its atoms vibrate more rapidly, and its temperature increases. **Heat** refers to the transfer of this energy between objects of different temperatures. For example, when you put a cold spoon into a mug of hot tea, the spoon receives energy from the tea and is heated.

The food that you eat every day, as well as the gasoline you put into your car's tank, has **chemical energy**. Chemical energy is actually an example of potential energy. Think about the food you ate today for breakfast. The energy from that food is released slowly into your blood stream throughout the morning. The actual bowl of oatmeal had potential energy that just happens to be in the form of chemical energy.

Perhaps the most important form of energy of all is **radiant energy**, or the energy of electromagnetic waves. Solar energy, the basis of most life on Earth, is radiant energy.

MEASURING ENERGY

Energy can be measured in terms of units of heat or as work. Units of heat are **calories**. One calorie is the amount of energy needed to heat one gram of pure water to one degree Celsius. A **joule** (J) is the work done when one kilogram is accelerated at one meter per second. A calorie is equal to 4.184 J.

ENERGY TRANSFERS AND THERMODYNAMICS

Energy is transferred in natural processes. The study of these transfers is **thermodynamics**. Thermodynamics provides an explanation for the rates of flow of energy and the transformation of energy from one form to another.

The basics of the laws of thermodynamics can be summarized as follows:

- The first law of thermodynamics states that energy is neither created nor destroyed as it moves through natural systems.

- The second law of thermodynamics states that energy is constantly degraded to lower forms as it is used. **Entropy**, the state of disorder, increases in natural systems. This means that there is less useful energy available when a process is completed than there was when that process began. Consequently, this means that everything in the universe is tending toward falling apart, slowing down, and becoming more disorganized.

One way to think of these laws and processes is to think about energy from the Sun. A constant supply of energy from the Sun is needed to keep biological processes going on Earth. Biological systems can use the energy as it flows through a system. The energy can also be temporarily stored in the system (e.g., in the chemical bonds of an organic molecule), but eventually that energy is released and dissipated into the environment. Thermodynamics offers an explanation and a model for this type of movement.

ENERGY CONSUMPTION: PAST, PRESENT, AND FUTURE

Growth of the human population over the past 300 years has been based on the availability of resources—most significantly, on the availability of cheap, abundant fossil fuels. Statistics show that the population of the United States grew from 76 million in 1900 to 294 million in 2004. The United States is growing faster than any other developed country and is showing no signs of leveling off. More than half of the 2.9 million people who were added to the population of the United States in 2004 were the result of more births than deaths. The rest came from immigration, both legal and illegal. The growing population of this country and others is putting an increasing strain on the natural resources of Earth.

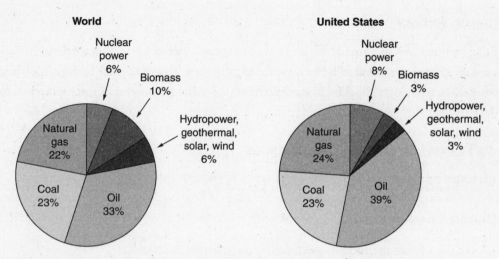

Figure 7.2 U.S. and World Commercial Energy Use in the Year 2002. In the year 2002, the United States used as much commercial energy as the rest of the world combined.

USE OF GLOBAL ENERGY

Approximately 99 percent of the energy that heats the surface of Earth, as well as heats our homes, comes from the Sun. Solar energy comes from fusion reactions within the Sun and is essentially an **inexhaustible** resource. The Sun is also indirectly responsible for other types of energy. These **renewable** sources of energy include wind power, hydropower, and biomass. Wind is created by differences in the heating of Earth's surface. The water cycle is powered by the Sun's energy as the Sun evaporates water from oceans, lakes, and rivers. And biomass is the conversion of solar energy to the chemical energy that is stored in the chemical bonds of organic material such as trees and other plants.

Commercial energy accounts for the other 1 percent of the energy used on Earth. Most of this commercial energy is the result of the extraction and burning of **nonrenewable resources**. Nearly 78 percent of the commercial energy used today comes from fossil fuels. This consists of natural gas, oil, and coal. Nuclear power is also considered to be a nonrenewable resource and accounts for approximately 6 percent of commercial energy. The rest comes from renewable sources such as hydropower, geothermal power (energy from internal sources of heat of Earth), solar energy, and wind as well as biomass.

FUTURE NEEDS

We will spend quite a bit of time in this chapter looking at renewable sources of energy. Sources such as solar energy, wind, geothermal power, and hydroelectric energy are potential alternatives to fossil fuels. Even though it is true that not one of these renewable energy sources will replace our dependence on fossil fuels and nuclear energy, it is possible that using several renewable sources of energy, with fossil fuels, may allow us to live in a sustainable, environmentally friendly way.

Scientists and economists predict that by the year 2100 it is possible that renewable energy sources could provide all our energy needs. Accomplishing this will involve the commitment of political leaders to pass laws that encourage conservation, protect the environment, and take the possibility of global warming very seriously.

FOSSIL FUEL RESOURCES AND USE

Fossil fuels are energy-rich substances formed from the remains of organisms. Petroleum, natural gas, and coal provide about 86 percent of the commercial energy in the world.

COAL

Coal is a solid fossil fuel formed from the remains of land plants. Most coal formed from plants that lived 300 to 400 million years ago. It is composed primarily of carbon but also contains small amounts of sulfur. When the sulfur is released into the atmosphere as a result of burning, it can form SO_2. SO_2 is a corrosive gas that can damage plants and animals. When it combines with H_2O in the atmosphere, it can form sulfuric acid, one of the main components of acid rain. Coal that is burned has also been shown to contain trace amounts of mercury and radioactive materials.

FORMATION OF COAL

Coal forms in stages (outlined in Figure 7.3). The type of coal that is being burned has an impact on the amount of sulfur that is released into the atmosphere.

The formation of coal goes through different stages as heat and pressure act on decomposing plant matter over millions of years. Coal begins as **peat**, partially decayed plant matter. Peat is found today in swamps and bogs, and it can be burned, but it produces little heat. As the decayed plant material is compressed over time, **lignite** is formed. Lignite is a sedimentary rock with low sulfur content and also produces a small amount of heat when burned.

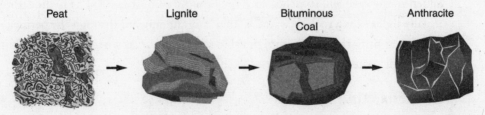

Figure 7.3 Stages of Coal Formation

Lignite becomes **bituminous coal**, the form of coal most widely used. It is also a sedimentary rock, but it has a high sulfur content. **Anthracite**, or hard coal, is a metamorphic rock formed when heat and pressure are added to bituminous coal. Anthracite coal is most desirable because it burns very hot and also contains a much smaller amount of sulfur, meaning that it burns cleaner. But the supplies of anthracite coal on Earth are limited.

EXTRACTING COAL

As discussed in Chapter 6, coal is gathered from mines. For subsurface mines, machines dig shafts and tunnels underground to allow the miners to remove the material. Buildup of poisonous gases, explosions, and collapses are all dangers that underground miners must face. The Sago Mine disaster of January 2006 in West Virginia—where only 1 of 13 trapped miners survived an explosion—shows how extracting these underground deposits of solid material is still a very dangerous process. Strip mining, or surface mining, is cheaper and less hazardous than underground mining. However, it often leaves the land scarred and unsuitable for other uses. (See Figure 7.4.)

PETROLEUM

Petroleum, or crude oil, is a thick liquid that contains organic compounds of hydrogen and carbon, called hydrocarbons. Crude oil can be separated into products, such as gasoline, heating oil, and asphalt. Petroleum contains many different types of hydrocarbons as well as sulfur, oxygen, and nitrogen.

FORMATION OF PETROLEUM DEPOSITS

Millions of years ago, organic material settled on the bottom of oceans. This organic material mixed with mud and was covered in sediment more quickly than it could decay. Thousands of years later, the sediments that contained the organic material were subjected to intense amounts of heat and pressure, changing it into a waxy material called kerogen. This substance is found in oil shales. When more heat was added to the kerogen, it melted into the liquid we know as oil. The hydrocarbons are usually lighter than rock or water, so they migrate upward through the permeable rock layers until they reach impermeable rocks. The areas where oil remains in the porous rocks are called reservoirs. (See Figure 7.5.)

EXTRACTING OIL

Oil is traditionally pumped out of the layer of reserves found under the surface of Earth. (See Figure 7.6.) An oil well is created using an oil rig, which turns a drill bit. After the

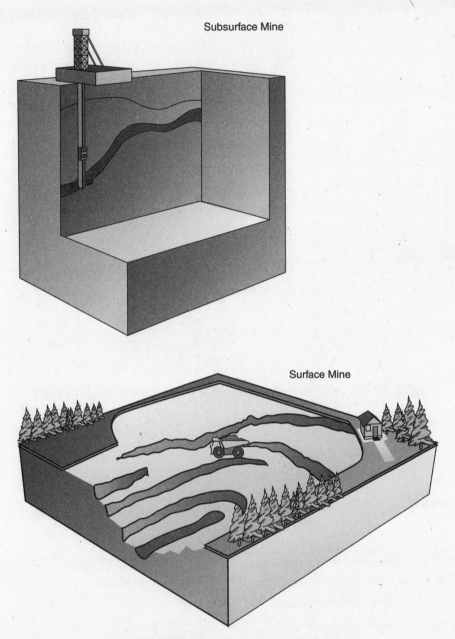

Subsurface Mine

Surface Mine

Figure 7.4 Surface and Subsurface Mining

hole is drilled, a casing—a metal pipe with a slightly smaller diameter than the hole—is inserted and bonded to its surroundings, usually with cement. This is done to strengthen the sides of the hole, or wellbore, and keeps dangerous pressure zones isolated. This process is repeated with smaller bits and thinner casings, going deeper into the surface to reach the reservoir. Drilling fluid is pushed through the casings to break up the rock in front of the bit. It also cleans away debris and rocks, and lowers the temperature of the bit, which grows very hot. Once the reservoir is reached, the top of the wellbore is usually equipped with a set of valves encased in a pyramidal iron cage called a Christmas Tree.

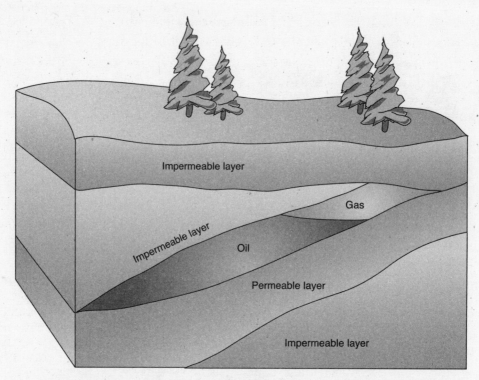

Figure 7.5 Oil and Earth's Impermeable and Permeable Rock Layers

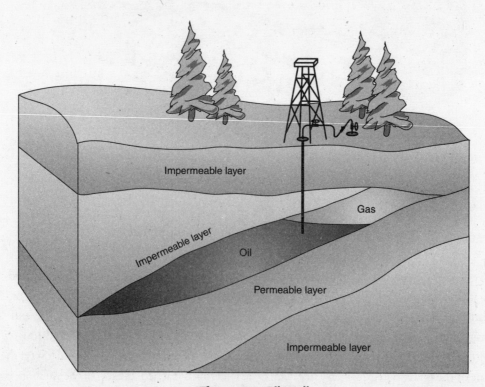

Figure 7.6 Oil Well

The natural pressure within the reservoir is usually high enough to push the oil or gas up to the surface. But sometimes, additional measures, called **secondary recovery**, are required. This is especially true in depleted fields. Installing thinner tubing is one solution, as are surface pumpjacks—the structures that look like horses repeatedly dipping their heads.

It is impossible to remove all of the oil in a single reservoir. In fact, a 30 percent to 40 percent yield is typical. However, technology has provided a few ways to increase drilling yield, including forcing water or steam into the rock to "push" out more of the oil. Even with this technique, only about 50 percent of the deposit will be extracted.

There are also more unconventional sources of oil, referred to as oil shale and tar sands. The hydrocarbons obtained from these sources require extensive processing to be useable, reducing their value. The extraction process also has a particularly large environmental footprint.

NATURAL GAS

Natural gas is the third most used commercial fuel. It provides approximately 23 percent of the energy consumed on Earth. Natural gas has many benefits over oil or coal. It is convenient, cheap, and produces only about half of the CO_2 of the other fossil fuels. But it is difficult to ship across the oceans and to store in large quantities. Natural gas found in remote areas is often simply burned off because pipelines and other methods of transporting the gas would be too expensive to make the fuel a useful resource. Estimates put the amount of natural gas wasted, or burned, at about 100 billion m^3 each year.

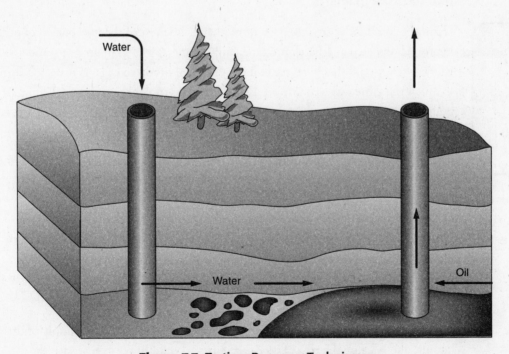

Figure 7.7 Tertiary Recovery Techniques

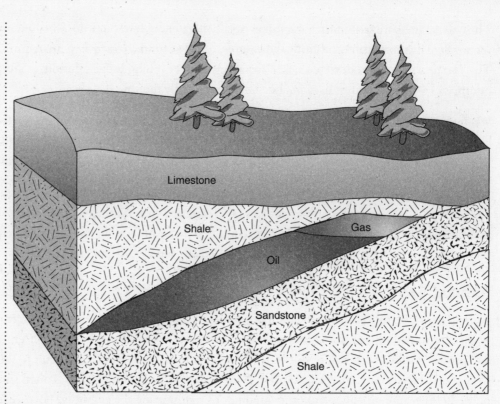

Figure 7.8 Natural Gas Deposit

NATURAL GAS DEPOSITS

Natural gas contains 50 to 90 percent methane. The rest of the gas consists of other hydrocarbons such as ethane, propane, and butane as well as hydrogen sulfide, which is a toxic gas that gives natural gas the smell of rotten eggs. Natural gas forms in much the same way as crude oil. Deposits of phytoplankton and small animals built up on the seafloor and were quickly buried by sediment. Most natural gas deposits sit atop a deposit of crude oil, capped by an impermeable rock layer. (See Figure 7.8.)

SYNFUELS

Synfuel, or synthetic fuel, is a liquid fuel obtained from coal or natural gas. An intermediate step in the production of synthetic fuel is often syngas, a mixture of carbon monoxide and hydrogen produced from coal that is sometimes directly used as an industrial fuel. Coal can be converted to synthetic natural gas by coal gasification or to a liquid fuel, as methanol, by the process of coal liquefaction.

Using synthetic fuels has many benefits. There is a potentially large supply of these liquids, and because the cost of converting them could prove to be relatively reasonable, the U.S. government has an interest in subsidizing such a process. Synfuels burn cleaner than coal, releasing less particulate matter into the atmosphere.

Although the particulate matter is reduced, current synfuels have higher CO_2 emissions than coal. There is a push with the U.S. Department of Energy and oil companies to reduce the CO_2 emissions during coal gasification. Also, there is a low to moderate net energy yield in many cases. The resulting synfuel will most likely cost more than coal. There is also a high environmental impact to creating this fuel because it demands the mining of at least 50 percent more coal. This will increase the amount of surface mining as well as the use of water in the mining process.

NUCLEAR ENERGY

Scientists and politicians alike predicted in the 1950s that nuclear power would provide clean and abundant energy and fill the gap created as the known supplies of oil and natural gas dwindled.

FISSION AND FUSION

Nuclear fission is the process of splitting the nucleus of an atom into two different and smaller particles. The fission reaction starts with a relatively large atom that has an unstable nucleus. A neutron is then directed at the nucleus at a high speed. When the neutron hits the nucleus of the large, unstable atom, the nucleus splits into two smaller nuclei and two or more neutrons.

The diagram in Figure 7.9 shows the fission reaction of an atom of Uranium-235. The nucleus of U-235 is hit with a neutron. The nucleus splits into two different atoms: Krypton-92 and Barium-141. More neutrons, as well as a great deal of energy, are also released. These neutrons can then strike other atoms, creating a chain of reactions. The important thing to remember here is that there is more and more energy released with each reaction. The energy is released as heat that

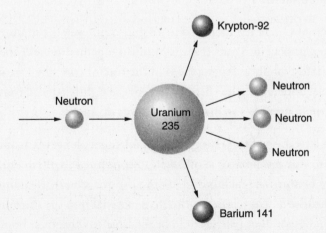

Figure 7.9   Nuclear Fission Reaction

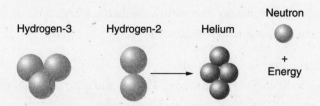

Figure 7.10 Nuclear Fusion Reaction

can then in turn be converted into electricity. This release of energy can lead to a huge explosion if the reactions are not controlled carefully.

Nuclear fusion is a second type of nuclear reaction. Nuclear fusion involves combining two atomic nuclei to form one bigger nucleus. (See Figure 7.10.) In this example, two nuclei of hydrogen (Hydrogen-2 and Hydrogen-3) are forced together by heat and pressure. The result is an atom of helium, a neutron, and energy.

One of the biggest obstacles in the way of creating fusion energy on Earth is temperature. Fusion occurs on the Sun but at temperatures of 15 million degrees C. Scientists have had some success creating the proper circumstances by using giant magnets, but, at the moment, the technology used to produce a fusion reaction is more expensive than the energy that is created.

Nuclear Fuel

Uranium-235 is a naturally occurring radioactive isotope of uranium and is the most commonly used fuel in nuclear power plants today. In most cases, U-235 makes up a very small percentage of uranium ore. Obtaining U-235 requires that the uranium be concentrated and purified through a series of chemical or mechanical processes. The mining and concentration of this uranium ore is often considered to be more hazardous than coal mining. Exposure to radioactive debris makes this a particularly dangerous profession.

Once the concentration of U-235 has increased to a significant level, the uranium is formed into pellets that are then packed into rods that are bundled together to make the **fuel assembly**. Thousands of these fuel assemblies are then placed in the reactor core, the location where the unstable uranium nuclei undergo nuclear fission. Due to the tightly packed rods of uranium, there is a chain reaction set up that releases tremendous amounts of energy.

These reactions and the creation of this energy can be controlled, however. Control rods are placed between the fuel rods to stop reactions or allow them to continue at a different rate. Some type of coolant fluid circulates between the fuel rods to remove the excess heat. A failure of this coolant system can lead to a meltdown that releases radioactive material into the surrounding environment. If this were to happen, thousands of people could be required to leave their homes, or lose farming lands, or could die of cancers, particularly thyroid cancer.

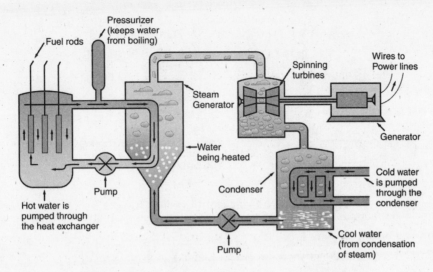

Figure 7.11 Pressurized Water Nuclear Reactor

TYPES OF REACTORS

Evaluating the costs and benefits of nuclear power requires exploring how a typical nuclear power plant operates. There are two main types of nuclear reactors: pressurized water reactors and boiling water reactors. As seen in the following descriptions, there are inherent advantages and disadvantages to either type of reactor and to nuclear power in general.

PRESSURIZED WATER REACTORS

As many as 70 percent of all the nuclear power plants in the United States and in the world are pressurized water reactors. (See Figure 7.11.)

In this design, water circulates through the core of the reactor, where it cools the fuel rods and absorbs that heat. The hot water is then superheated to 317°C and reaches a pressure of 2,235 psi. The water is pumped to a steam generator where it heats another section of the reactor that is also filled with water. In this second section, steam is created that drives a high-speed turbine that in turn produces energy.

There is waste heat generated during this process. Some plants use huge cooling towers, illustrated in Figure 7.12, to release the heat into the atmosphere.

BOILING WATER REACTOR

The boiling water reactor is simpler, but it can be dirtier and more dangerous than the pressurized water reactor. Water from the reactor boils, making steam, which drives turbine generators. The concern is that, in this model, radioactive water and steam is no longer confined to the containment structure. It can be difficult to contain and control leaks, and the chances of releasing the radioactive water or steam are pretty high.

Figure 7.12 Nuclear Plant Cooling Towers

CHERNOBYL

Many reactors in Europe, including the doomed Chernobyl nuclear power plant in Ukraine, use graphite as the material that encases the core. Scientists initially assumed that these designs were very safe because graphite has a high capacity for capturing neutrons and dissipating heat, but it is important to realize that graphite will burn in the presence of air. A fire in the graphite core in the reactor in Chernobyl allowed nuclear fuel to melt and escape from the reactor.

PROS AND CONS OF USING NUCLEAR POWER

Politicians, environmentalists, and activists alike have argued the pros and cons of nuclear energy for decades. Costs, storing the wastes, and vulnerability to terrorist attacks have sent both sides of the argument scrambling to support their positions.

PROS

When evaluating the costs and risks of nuclear fuel, look at the big picture. The entire process involves the following: the steps needed to mine the uranium used to make nuclear fuel; the steps involved in processing it to make the fuel; the use of it in a nuclear reactor; the need to safely store the radioactive waste products; and the potential for dealing with a radioactive reactor that has exceeded its useful life.

The advantages to the process are actually many. Nuclear power creates a large supply of commercial fuel. If the plant can run without an accident, there is a low environmental impact with moderate land disruption and water pollution. Many contend that there is a low risk of exposure to radioactivity because of a series of multiple built-in safety features and systems that are in place. Many believe that an increase in the nuclear power capabilities in the United States will reduce the dependency on foreign oil. The use of nuclear power would also reduce the amount of carbon dioxide emitted into the atmosphere because nuclear power releases one-sixth as much CO_2 as coal.

Cons

For every argument of why nuclear power should be used, there is a counterpoint arguing that it is not a viable fuel source. Opponents point to the high cost of building and maintaining a nuclear facility, even with the anticipated government subsidies. The environmental impact of an accidental meltdown or partial meltdown could be catastrophic. The nuclear power plants are also potential sites for terrorist attacks. Many people also point out that energy facilities such as these spread the knowledge and the technology that could be used to build nuclear weapons. The process of making power from a nuclear fuel creates a highly radioactive waste. Many argue that there is no widely acceptable plan or solution for handling, transporting, and storing this waste or for decommissioning a highly radioactive nuclear power plant.

Radioactive Waste

In addition to high-level radioactive waste, there are low-level radioactive wastes generated that also need to be disposed of properly. A low-level radioactive waste has been categorized as a material that gives off a small amount of radiation and needs to be stored for 100 to 500 years before it has decayed to safe levels. Things like tools, building materials, clothing, and glassware can be contaminated by radioactivity and be considered low-level wastes. Low-level radioactive wastes can be generated in nuclear power plants, hospitals, universities, and other industries. Today, low-level radioactive wastes are placed into steel drums and transported to regional landfills run by state and federal governments.

High-level radioactive wastes are not so clear cut. These radioactive wastes release large amounts of harmful radiation for a short time and then release a small amount of radiation for a very long time. This means that a high-level radioactive waste needs to be stored for 10,000 to 240,000 years. Some believe that it is possible to safely contain such waste over that period of time while others say that it is impossible to ensure this.

Following are possible methods for dealing with high-level radiation along with the argument against it:

- Bury it deep underground: *It is difficult to predict which sites will be geologically stable for such a long period of time.*

- Shoot it into space or toward the Sun: *This is very expensive, and a disaster at the launch of the rocket could prove to be devastating.*

- Bury it under an ice sheet in Antarctica or Greenland: *The long-term stability of these ice sheets is not well known, especially because the heat from the radioactive material could hasten melting. This practice is prohibited by international law.*

> **AP EXPERT TIP**
>
> Regardless of the actual problems or merits of nuclear power, one obstacle to new nuclear plants is the "NIMBY" (Not In My Backyard) issue. Essentially, NIMBY means that people don't mind a necessary but unpleasant facility (nuclear power plant, wind farm, recycling plant, landfill, etc.) as long as it is located somewhere else. While this attitude is understandable, if everyone takes this approach some tasks can be difficult or impossible to complete.

- Dump it into subduction zones in the ocean: *The waste could eventually be released in a volcanic eruption and the containers contaminate the ocean. There are international laws that also prohibit such a solution.*

- Bury it in thick deposits of shale in the deep-ocean: *The containers would eventually corrode and release the radioactive material. This is also not allowed by international law.*

- Change it to less harmful isotopes: *There is currently no way to do this.*

HYDROELECTRIC POWER

Hydropower is a renewable resource that can be harvested and used to generate electricity. Many countries today generate a significant amount of their electricity from moving water. Norway, for example, produces nearly 99 percent of its electricity from moving water.

DAMS

It is possible to harness the energy of water flowing over a waterfall. But most of the hydroelectric power in the world has been generated with the use of dams in recent years. A dam creates a boundary on a river, builds up a lake behind, and stores the water to be released to generate electricity. Dams are fairly controversial structures, having positive and negative affects on the environment and local surroundings. The main purpose of a dam is to produce cheap electricity though the use of hydropower. (See Figure 7.13.) Dams can also reduce flooding in downstream

Figure 7.13 Typical Dam

areas and provide water for irrigation. They also create large water reservoirs that can be used for recreation and fishing.

CONTROLLING FLOODS

Building a dam reduces the threat of flooding by storing water in a reservoir and allowing the release of water to be controlled. The Three Gorges Dam in China on the Yangtze River, which has a history of extreme flooding, is slated to be completed in 2013. Estimates place the number of people who have died in flood events in the past 100 years to be more than 500,000 people, with 4,000 of them casualties of a 1998 flood. This dam, and others built elsewhere, could prevent many of these disasters.

THE ENVIRONMENTAL IMPACTS OF DAMS

Many argue that the benefits of building a dam along a river do not outweigh the harm. Dams significantly alter the landscape upstream from the dam. Look back at Figure 7.13. Significant amounts of water are lost from the reservoir due to evaporation, especially in warm climates. Some dams even lose more water to evaporation and seeping than they store. The land adjacent to the river upstream from the dam will be flooded, resulting in the loss of habitat, farmlands, archaeological sites, and homes. The migration and spawning of fish, such as salmon, will be disrupted after the dam is built. Sediment flowing in the river will build up in the reservoir and become troublesome for a number of reasons. First, the reservoir behind the dam will eventually fill with sediment that will make the dam useless for spring water or producing electricity. Second, the sediment will need to be removed, which is a costly exercise. Third, the buildup of silt behind the dam deprives the downstream regions of fertile, nutrient-rich soil that would otherwise naturally replenish the area when the river floods.

SOLAR ENERGY

The practice of gathering and holding heat from the Sun by using objects with no moving parts is called **passive heat absorption**. Consider adobe structures—the thick walls would slowly collect the heat during the day and then slowly release it at night. This allowed the residents to maintain a comfortable daytime temperature indoors. Building a greenhouse on the southern side of a building today is an example of using passive heat. Modern solar power technology collects energy from the Sun with solar panels, thin sheets consisting of **photovoltaic cells**, typically made of semi-conducting silicon. These photovoltaic cells convert sunlight directly into electricity. These require a lot of surface area for maximum efficiency. Another commercial solar collecting method, **concentrating solar power**, uses curved mirrors to focus sunlight to heat a liquid. The heated liquid is then used to run a conventional power plant, instead of liquid heated by coal or oil.

Active heat systems pump a heat absorbing fluid (such as water, air, or antifreeze) through a small collector. These collectors are usually placed near or on top of buildings rather than being built into them as is done with a passive solar heating system.

The biggest challenge with solar energy is storing the heat for times when there is no sunlight. One inexpensive solution uses small, insulated water tanks to store heat generated by the Sun shining on them. This system works best in areas where there are few days without sun and where the daily temperature does not vary much over the course of the year.

WIND ENERGY

Wind is another form of solar energy. Wind is created by the unequal heating of Earth's surface by the Sun and by the rotation of Earth on its axis. This indirect source of solar energy can be captured by wind turbines and then converted to electricity.

Wind energy is very popular in Europe. In fact, about 75 percent of the world's wind power is produced in inland and offshore wind farms in Europe. In the United States, a growing number of farmers are leasing their land for wind power. (See Figure 7.14.)

Experts feel that the advantages to wind power outweigh the disadvantages. Wind power does not create CO_2 emissions. Many people contend that wind power has very little environmental impact. In many areas, the strongest winds blow over the ocean, so many current and planned wind farms are situated far offshore. Consider also the fact that land beneath wind turbines can still be used for

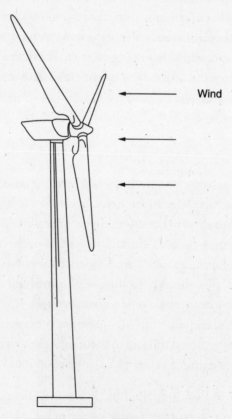

Wind

Figure 7.14 Wind Turbine

farming or grazing livestock. Wind farms are quickly and easily constructed and expanded. It seems like a good alternative, at least as a backup electricity source.

Some people disagree. They point to the fact that wind farms are only useful in areas of steady winds and that backup systems are needed in times of quiet winds. Many have concerns about the pollution created by wind farms—visual and noise pollution. Other people worry about the impact that large wind farms will have on migratory bird routes or the dangers birds or bats may experience with the wind drafts created around the wind turbines.

THE OCEAN AS AN ENERGY SOURCE

The energy contained in ocean tides and waves is enormous and can be harnessed to do useful work. A **tidal station** works much like a hydropower dam. (See Figure 7.15.) The turbines spin as the tide flows over them. This turbine is connected to a generator that produces electricity. The incoming and outgoing tides are held back by a dam in a tidal station. The difference in water levels generates electricity in both directions as the water runs through a series of reversible turbines. Of course, this type of system does depend on the tides. Because the average tidal period is 13.5 hours, the timing of the high and low tides rarely coincides with peak electricity usage hours. Even so, the demand for this type of renewable and constant energy source is large in certain regions around the world.

Technology is also currently being developed to effectively harness ocean waves as an energy source. The energy that waves expend as millions of tons of water are moved each day far surpasses the energy generated by solar and wind power in specific areas. If this energy were captured and converted to commercial use, it would be a significant contribution to localized energy needs. Scotland opened the first commercial wave-power station in 2001. This facility currently generates enough electricity for 400 homes.

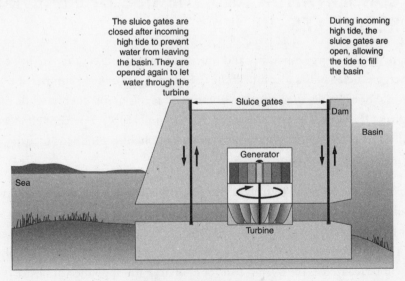

Figure 7.15 Tidal Station

GEOTHERMAL ENERGY

We could also use the energy stored in Earth's mantle to heat and cool buildings and to generate electricity. **Geothermal energy** is the heat found in soil, rocks in the ground, and fluids in Earth's mantle. Yosemite National Park is the largest geothermal region in the United States. Iceland, Japan, and New Zealand have high concentrations of geothermal vents and springs. These sources can produce wet steam, dry steam, or hot water. In most places on Earth, the temperature at a depth of about 3 m is 10°C to 16°C.

For a long time, geothermal energy was used for therapeutic baths built at hot springs. More recently, this energy has been harnessed to generate electricity for industry, space heating, agriculture, and aquaculture. The technology exists (pumps and pipes) to use the difference in temperature between the surface and the subsurface to heat or cool a building. Quite a few advantages accompany the use of geothermal energy to generate electricity. Geothermal generators have a relatively long life span. Little waste needs to be disposed of from this process, and no mining or transportation costs factor into the equation. There is moderate environmental impact, as it requires low land use and disturbance.

Critics of geothermal power cite the scarcity of available and suitable sites as a leading disadvantage to geothermal power. They point out that this is a source that can be depleted if it is used too rapidly. There are some CO_2 emissions. There is some moderate to high air pollution close to the site and a noise and odor (H_2S) problem. These people point out that the cost of this technology is too high in areas other than those at the most concentrated and accessible sites. Sites must be chosen very carefully, as careless site selection has been proven to cause earthquakes.

ENERGY CONSERVATION AND THE USE OF RENEWABLE ENERGY RESOURCES

It is important to discuss energy efficiency and the use of energy sources that are not derived from fossil fuels. **Energy efficiency** is a measure of the useful energy produced compared to the energy that is consumed. For example, the light that a bulb produces is useful energy, while the heat that the bulb gives off is wasted energy. An incandescent bulb is only 5 percent efficient, while a fluorescent bulb is 22 percent efficient; the same amount of light is produced by using roughly one-quarter of the energy. (See Figure 7.16.)

Some people feel that the only way to truly conserve energy is to create new technology to replace existing technology, as in the example of the lightbulb in Figure 7.16. Individuals can save energy, and money, by buying more energy-efficient cars, heating systems, appliances, lights, and air conditioners. It is true that many of these energy-saving devices may cost more initially, but an overall saving will be realized as the overall cost for operating the appliance or device is lower.

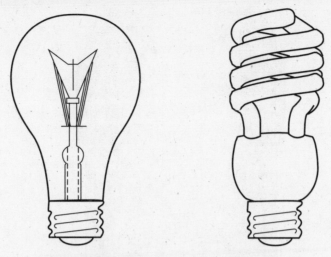

Figure 7.16 Lightbulb Energy Efficiencies

The automobile is a widely used device that wastes a large amount of energy. Automobiles with internal combustion engines waste about 75 percent to 80 percent of the energy in their fuel. Standards set up by the Environmental Protection Agency (EPA) have sought to regulate such waste. In 1975, Congress enacted the Energy Conservation Act. The Corporate Average Fuel Economy (CAFE) standards were added to this act as an amendment. The CAFE standards established fuel efficiency standards for light trucks and cars. CAFE standards are the sales-weighted average fuel economy, in miles per gallon, of a vehicle weighing 8,500 pounds or less. As a result of these CAFE standards, between 1973 and 1985, the average fuel efficiency rose for all cars sold in the United States. But from 1985 to 2004, the fuel efficiency for new cars leveled off or declined slightly as the size and power of cars increased.

HYBRID VEHICLES

Vehicles with a gasoline-electric engine have the highest efficiency and the lowest emissions levels of any vehicle available in the United States. These fuel efficient hybrid cars are powered by a battery that is recharged by energy from braking and a small internal combustion engine. The car still needs gasoline, diesel fuel, or natural gas to operate, and the battery is used for acceleration and for climbing hills. (See Figure 7.17.)

HYDROGEN FUEL CELLS

Technology now exists to generate energy locally rather than having to depend on storing and transporting that energy. A **fuel cell** is a device that uses an ongoing electrochemical reaction to generate an electric current. It is like a battery, but instead of recharging the cell with an electric current, you add more fuel for the chemical reaction. Essentially, a fuel cell converts hydrogen and oxygen into water and, in the process, produces electricity. (See Figure 7.18.)

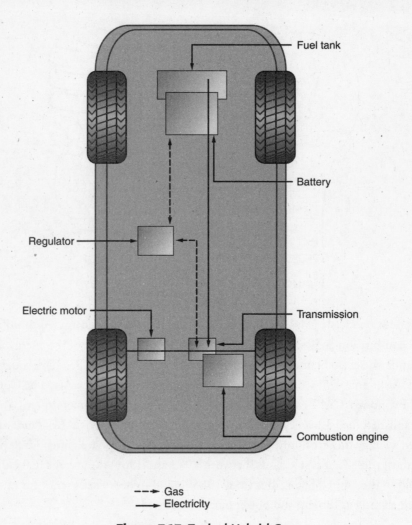

Figure 7.17 Typical Hybrid Car

Many people believe that in the future the fuel cell will compete with many other types of energy conversion devices. The fuel cell could someday replace the battery in a laptop computer or the engine in an automobile. Combustion engines, such as the turbine and the gasoline engine, burn fuels and use the pressure created by the expansion of the gases to do mechanical work. Batteries convert chemical energy back into electrical energy when needed. Eventually, fuel cells should do both tasks more efficiently.

There are several different types of fuel cells, each using a different chemistry. Fuel cells are usually classified by the type of electrolyte they use. Some types of fuel cells work well for use in stationary power generation plants. Others may be useful for small portable applications or for powering cars.

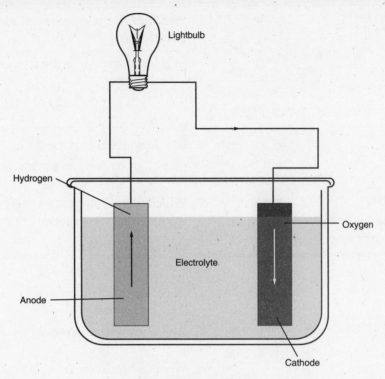

Figure 7.18 Fuel Cell

As seen in Figure 7.18, a fuel cell has a positive electrode (cathode) and a negative electrode (anode). These are separated by an electrolyte, a material that allows ions, but not electrons, to pass through. In most fuel cells, hydrogen and oxygen are used. These fuel cells create no waste products; they create radiant heat and water that is clean enough to drink. The efficiency of the fuel cell is offset somewhat by the large amounts of energy required to produce hydrogen.

REVIEW QUESTIONS

1. Energy is usually released in which of the following forms?

   (A) Heat
   (B) Light
   (C) Work
   (D) Wastes
   (E) Water

2. Explosions and collapses are dangers in the mining of which fossil fuel?

   (A) Natural gas
   (B) Coal
   (C) Petroleum
   (D) Geothermal energy
   (E) Uranium

3. Which of the following forms of coal is a metamorphic rock?

   (A) Peat
   (B) Lignite
   (C) Bituminous coal
   (D) Anthracite
   (E) Petroleum

4. Synfuels are produced from the alteration of which energy source?

   (A) Hydrogen
   (B) Petroleum
   (C) Hydroelectric power
   (D) Geothermal
   (E) Coal

5. Which of the following is an advantage of nuclear energy?

   (A) Emits one-sixth the amount of CO_2 as does coal
   (B) Open to the threat of terrorist attacks
   (C) Possibility of accidents
   (D) Low net energy yield
   (E) High costs to set up and maintain

6. Which type of alternative energy source has the potential for visual and noise pollution?

   (A) Geothermal energy
   (B) Hydroelectric energy
   (C) Wind power
   (D) Tidal power
   (E) Ocean-wave power

7. What environmental impact can a dam have on the area downstream from the dam?

   (A) The land is flooded, destroying homes and crops.
   (B) Large amounts of water are lost through evaporation.
   (C) Flooding is reduced.
   (D) Sediment builds up.
   (E) An area is created for fishing and recreation.

8. Which of the following devices use electrochemical reactions to produce an electric current?

   (A) Battery
   (B) Fuel cell
   (C) Synfuel
   (D) Solar panels
   (E) Fuel rods

9. Which is the most feasible solution to storing high-level radioactive waste?

   (A) Burying it under the ice sheets in Antarctica

   (B) Dumping it into a subduction zone

   (C) Burying it deep underground

   (D) Sending it off into space

   (E) Altering the radioactive elements

10. Which fuel is wasted because it is too expensive and difficult to transport long distances?

   (A) Petroleum

   (B) Nuclear

   (C) Gasoline

   (D) Natural gas

   (E) Coal

FREE-RESPONSE QUESTION

Alternatives to fossil fuels are certainly needed for many reasons, but the most pressing environmental reason is to reduce the global warming gases in the atmosphere.

(A) (4 points) Name two greenhouse gases and connect each with a source.

(B) (2 points) Explain a short-term health benefit of reducing air pollutants.

(C) (3 points) Identify three energy sources that are alternatives to fossil fuels.

(D) (2 points) Name and explain one piece of legislation or an international treaty that deals with air pollution.

ANSWERS AND EXPLANATIONS

1. A

Energy is the ability to do work, so this eliminates choice (C). Energy does not release light or water, so this eliminates (B) and (E). It is true that many times energy does release some sort of waste product, as choice (D) suggests, but this is not an accurate way to describe the process.

2. B

Geothermal energy and uranium are not fossil fuels. This eliminates choices (D) and (E) immediately, even though uranium must be mined. Natural gas and petroleum are usually extracted with a well, not in a mine, leaving coal as the correct choice.

3. D

Petroleum is a fossil fuel, but it is not a form of coal. Peat is loosely compacted organic matter. Lignite and bituminous coal are sedimentary rocks. Anthracite is the metamorphic form of bituminous coal.

4. E

Synfuels are formed from the conversion of coal or natural gas. Hydrogen and petroleum cannot be altered to produce this type of synthetic fuel. Hydroelectric power and geothermal energy are both renewable resources that don't produce a definite, liquid product.

5. A

While nuclear fuel emits one-sixth the amount of CO_2 as does coal, (A), its low net energy yield, (D), is a detriment, not an advantage. The nuclear reactors are costly to set up and maintain, (E), and are open to the threat of terrorist attacks, (B), and the possibility of accidents, (C).

6. C

Geothermal energy may create a strange odor, but not much concern has been voiced about the visual and noise factor. The dams used for hydroelectric power are generally not loud and are frequently used as a recreation area. Tidal power and ocean waves do not create visual and noise pollution. This leaves the correct choice, wind power.

7. C

When a dam is built, the area upstream is flooded. This eliminates choice (A). Because there is also much evaporation from that reservoir, (B), is incorrect, too. Sediment builds up in the reservoir, and the reservoir is also used for fishing. This leaves choice (C) as the only factor that influences the area downstream of the dam.

8. B

A battery is basically the opposite of this, so choice (A) is incorrect. Synfuels, (C), are synthetic fuels produced from coal. Solar panels and fuel rods do not use electrochemical reactions, so this eliminates choices (D) and (E) as possible choices.

9. C

Choices (A) and (B) are prohibited by international law. The concern about shooting the waste into space is the expense and the possibility of a catastrophic launch. And there is no technology today that can change a radioactive element into one that is less harmful. Note that burying waste deep underground (C) is not risk-free, but it is the most realistic of the options presented.

10. D

Petroleum, gasoline, and coal are all very easily transported by truck or train. Nuclear energy is not transported great distances. But a large amount of natural gas is burned off every year in areas where it is isolated from pipelines. It is very expensive and difficult to transport natural gas.

ANSWER TO FREE-RESPONSE QUESTION

(A) Carbon dioxide and carbon monoxide are products of the combustion of fossil fuels. Nitrogen oxides are products of the combustion of fossil fuels. Other choices: methane from rice paddies, landfills, and cow flatulence, ozone linked to combustion of fossil fuels, and water vapor from many sources.

(B) Asthma is a huge and growing problem, especially for children in inner cities. Reducing air pollutants, including particulate matter and sulfur dioxide, would have a great impact on this problem. Diseases such as West Nile virus and other insect-borne diseases are also on the rise, the result of warmer temperatures that allow certain insects to populate areas from which they were previously absent. Limiting greenhouse gas emissions may slow the pace of climate change, and presumably reduce the range expansion of diseases.

(C) Any of the following will work: hydropower, solar power, wave energy, tidal energy, geothermal, biomass fuel, geothermal energy, nuclear energy.

(D) The Clean Air Act of 1972, updated in 1977 and 1990, and with standards that are constantly revised, is the most important piece of U.S. legislation that deals with clean air. This act enables the Environmental Protection Agency to set limits on specific air pollutants permitted in the United States. The Kyoto Protocol, the legally binding international agreement signed by over 100 countries, contains rules on reducing greenhouse emissions. The Energy Policy Act governs energy policy in the United States, including efficiency standards and alternative fuel research.

CHAPTER 8: POLLUTION AND ITS EFFECTS

IF YOU LEARN ONLY FIVE THINGS IN THIS CHAPTER . . .

1. Air pollution can occur outdoors or in homes and offices. In some cases, air quality inside is poorer than air quality outside.

2. The U.S. Congress passed the Clean Air Act in 1963. This act outlined seven major pollutants: sulfur dioxide, carbon monoxide, particulate matter, hydrocarbons, nitrogen oxides, photochemical oxidants, and lead.

3. Noise pollution is an example of pollution that can reduce the quality of life. Noise, odors, and light are examples of nuisances that can degrade your environment.

4. Any chemical, biological, or physical change in water quality that has a harmful impact on living organisms is considered to be water pollution. Water pollution can be split into two main categories: pollutants that can cause health problems and pollutants that cause disruption of the ecosystem.

5. The Clean Water Act, passed in 1972, sought to ensure that the surface waters in the United States are returned to swimmable or fishable states.

Pollution is the release of chemical, physical, biological, or radioactive contaminants or hazards into the environment. Among the most common forms of pollution are air, noise, and water pollution. Pollution is a tremendous problem around the world, mainly because it is usually what economists call an **externality**. An externality is a cost that is not included in the price of a good, but is paid for somewhere else. The environmental damage of pollution is a real cost to society, but is generally not reflected in the price of goods or production. One of the best ways to limit pollution is to internalize the cost by putting a price on pollution. This can be done through fees, taxes, or direct limits on emissions, although the last method does not take costs or benefits into account and is considered economically inefficient.

AIR POLLUTION

Air pollution is the release of chemical, physical, biological, or radioactive contaminants into the atmosphere. Most air pollutants are located in the troposphere, but some find their way to the stratosphere. Some of these pollutants can cause depletion of the stratospheric ozone layer, which will be discussed in Chapter 9. **Primary pollutants** are directly released from their source into the air from a specific location of highly concentrated pollution. A good example of this is a smoke stack emitting a dirty plume of ash and dust into the air. **Secondary pollutants** are transformed into hazardous material because of chemical reactions in the air. Radiation from the Sun often drives these chemical reactions.

SOURCES OF AIR POLLUTION

The sources of air pollution from human activity are usually related to burning different kinds of fuel. Some examples include combustion-fired power plants, motor vehicles, marine vessels, crop waste burning, solvent fumes, and methane from landfills, among others.

Some scientists talk of air pollution as being a problem caused entirely by humans, but there are also many natural causes. For example, ash, acid, hydrogen sulfide, and other toxic gases are released during volcanic eruptions. If the ash and gases reach high into the atmosphere, they may travel great distances via air currents. Sea spray and decaying vegetation release sulfur compounds as well. Trees and other plants release millions of tons of volatile organic compounds. Pollen, spores, viruses, and bacteria are just a few examples of organic materials in the air that can cause allergies and infections. Dust storms can also kick up dust and fine particles, carrying them to distant places.

Pollutants that are caused by dust and soil erosion, construction, or equipment leaks are known as **fugitive emissions** because they escape accidentally from their sources into the atmosphere. The amount of dust released as fugitive emissions each year in the United States nearly equals the amount of CO_2 released as fossil fuel is burned.

TYPES OF POLLUTION

The Clean Air Act of 1972 identified seven major pollutants: sulfur dioxide, carbon monoxide, particulate matter, hydrocarbons, nitrogen oxides, photochemical oxidants, and lead.

SULFUR DIOXIDE

Sulfur dioxide (SO_2) is a colorless corrosive gas that harms plants and animals. SO_2 can cause breathing problems in healthy people and restrict the airways of people with asthma. Exposure to SO_2 over an extended period of time can lead to a condition that is similar to bronchitis. SO_2 in the atmosphere reduces visibility and damages aquatic life in lakes. SO_2 can corrode metals and damage paints and leather. Materials containing calcite, such as the marble facades of buildings, are especially vulnerable. SO_2 may change to sulfur trioxide (SO_3) when it enters the atmosphere. SO_3, in turn, can react with water vapor to form sulfuric acid (H_2SO_4), which is the main component of **acid rain**. Sulfate ions (SO_4^{-2}) can be carried long distances or inhaled by animals, leading to lung problems. Some medical professionals assert that sulfur dioxide and sulfate ions are the second major cause of health issues related to air pollution. (Smoking is the first major cause.)

Nearly two-thirds of sulfur in the atmosphere is caused by human activities. The burning of fossil fuels such as coal and oil are two major causes. Also, some natural gas and oil sources must be cleaned of sulfur before they can be used, releasing more sulfur in the process. Countries that burn large amounts of coal, such as China and the United States, contribute significantly to the release of sulfur into the atmosphere.

Carbon Monoxide

Though colorless and odorless, carbon monoxide (CO) is a highly toxic gas that is formed by the incomplete combustion of fuel such as coal, oil, natural gas, or charcoal. CO binds to hemoglobin in the blood of animals, restricting respiration. Internal combustion engines, fires used to clear land, and cooking fires are major contributors of CO in the lower atmosphere. Nearly all the CO in the atmosphere eventually leads to the production of ozone. Ozone in the lower atmosphere is a major component in photochemical smog.

Particulate Matter

Dust, ash, soot, lint, smoke, pollen, and spores are all considered **particulate matter**, that is, solid or liquid particles suspended in a gas. Particulate matter is often the most obvious form of air pollution; it often reduces visibility and can leave windows and cars dirty. Particles that are less than 2.5 mm in size are among the most dangerous because they can directly enter the lungs, damaging lung tissue. Some of the most dangerous particles of this size are cigarette smoke and asbestos. Not only can these particles physically damage the tissues in the lungs, but they are also carcinogenic, meaning that they cause or promote cancer.

Nitrogen Oxides

Nitrogen oxides are highly reactive gases that are formed when nitrogen is heated to about 650°C. This can happen during the burning of fossil fuels or other sources of biomass, or lightning can cause it. Bacteria in the soil can also form nitrogen oxides. Nitrogen oxide (NO) is the initial gas formed when these events take place. This gas oxidizes further in the atmosphere, forming nitrogen dioxide (NO_2), which is responsible for giving photochemical smog its distinctive reddish brown color. Nitrogen oxides may also mix with water in the atmosphere, creating nitric acid (HNO_3), which is a major component in acid precipitation.

PHOTOCHEMICAL OXIDANTS

Photochemical oxidants in the atmosphere are products of a secondary reaction or of a reaction caused by the Sun. The creation of **ozone** is the most significant photochemical reaction; the reaction takes place when an atmospheric reaction splits a molecule of nitrogen dioxide, forming a single atom of oxygen. This single atom can then react with a molecule of O_2 to form O_3, or ozone. Ozone in the stratosphere protects the planet from the Sun's harmful radiation. Ozone that is formed close to the ground, however, can damage plants and building materials such as paint, rubber, and plastics, and can even damage eyes and lungs. Ozone gives photochemical smog its distinctive odor.

LEAD

Most of the lead in the atmosphere resulted from the burning of leaded gasoline. Because of its extremely hazardous nature, one of the most successful pollution control measures in U.S. history banned leaded gasoline in 1986 and has had amazing results. For example, the average lead levels in children's blood have dropped 90 percent. But the worldwide lead emissions still are at levels of nearly 2 million metric tons per year. Most of this comes from leaded gasoline burned in developing countries. To date, about 50 nations have joined the United States in banning leaded gasoline, so these levels should begin to decline.

TOPOGRAPHY AND AIR POLLUTION

Having discussed the major types of air pollution, we now need to explore why air pollution is more prevalent in some areas than in others. Topography, climates, and physical processes in the atmosphere also play an important role in transporting, concentrating, and mixing air pollutants.

TEMPERATURE INVERSIONS

Earth's surface absorbs heat, so under normal circumstances, temperature decreases as elevation increases. Even at the equator, which receives constant, intense solar radiation, many mountains and volcanoes are snow-capped. Convection currents, as seen in Figure 8.1, interrupt this normal decrease in temperature, causing **temperature inversions**, a stable layer of warm air above a layer of cooler air.

Los Angeles, California, is a good example of an area that often experiences temperature inversions. The city is surrounded by mountains on three sides and has a dry, sunny climate. The ground tends to cool very quickly at night, trapping heavier, cooler air (and any pollutants it contains) below the lighter, relatively warm air at higher altitudes. This prevents mixing of the air, leading to a concentration of pollutants released by cars and trucks during the day. Add the bright morning sunlight to this mix and you have a recipe for photochemical smog that can often be seen along the city's skyline. The Sun warms the land, allowing for convection currents to form again and pollutants concentrated during the night to move toward the ground. Even more contaminants

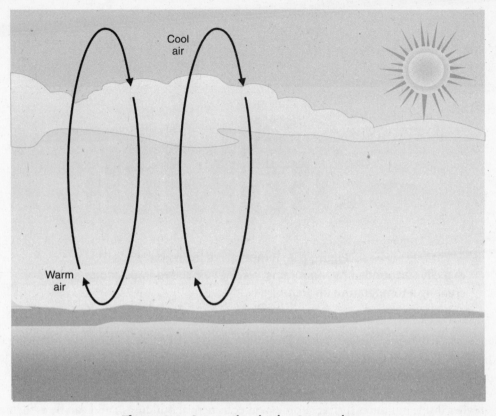

Figure 8.1 Convection in the Atmosphere

are added near the ground from automobiles, airplanes, lawn mowers, and the like, and the vicious cycle continues. (See Figure 8.2.) In addition, the surrounding mountains prevent the pollutants from dispersing, keeping them concentrated over the city, where they contribute to downdrafts.

HEAT ISLANDS

Concrete and glass in an urban area, coupled with the relative lack of vegetation, prevent rainwater from infiltrating the ground. This creates an abundance of runoff and high rates of heat absorption during the day and radiation at night. In addition, the tall buildings create updrafts, causing pollutants to be swept up into the air. Temperatures in large cities can be up to 5°C higher than the surrounding areas (**urban heat island**). All of these factors combined create a stable air mass that traps pollutants close to the ground.

INDOOR AIR POLLUTION

In many cases, the air we breathe indoors poses a greater threat to our health than the outside air. Over the past few decades, considerable effort has gone into regulating and controlling outdoor air pollution. Only in relatively recent times have people begun to realize that indoor air can be very dangerous.

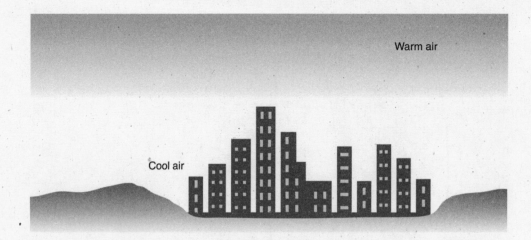

Figure 8.2 Temperature Inversion
In a city surrounded by mountains, warmer air above traps cooler air below, creating a temperature inversion.

Studies conducted by the Environmental Protection Agency (EPA) have found the following three alarming facts about indoor air pollution:

1. Levels of nearly all of the 11 common pollutants studied are typically 2 to 5 five times higher inside buildings (private homes and commercial buildings), and can be as high as 100 times higher in extreme cases.

2. Pollution inside automobiles, especially in areas with heavy traffic, can be up to 18 times higher than outdoors.

3. People typically spend 70 percent to 98 percent of their time indoors, accelerating the negative affects of indoor air pollution.

Figure 8.3 shows some major sources of indoor air pollution.

SMOKING

Tobacco smoke is the most significant air pollutant in the United States in terms of its impact on human health. It is estimated that smoking and its related diseases cause 20 percent of the deaths in the United States each year. Recently, however, smoking restrictions have resulted in dramatic declines in secondhand smoke exposure. Indoor smoking bans have also significantly decreased the levels of tobacco by-products in nonsmokers' systems.

FORMALDEHYDE

Formaldehyde, a colorless gas, is found in many common materials such as furniture stuffing, paneling, particleboard, and foam insulation, and it can cause a host of health problems. People exposed to formaldehyde may experience breathing problems, dizziness, rashes, headaches, sore

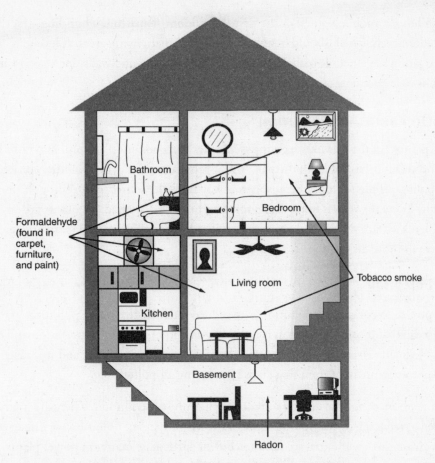

Figure 8.3 Major Indoor Air Pollution Sources

throats, eye irritation, or nausea. It is estimated that 1 in every 5,000 people who live in homes containing formaldehyde products for more than 10 years will develop cancer from their exposure.

Radon

Radon-222 is a naturally occurring, odorless, invisible gas that is produced by the decay of Uranium-238. Many rocks and minerals contain a small amount of Uranium-238, though underground deposits of materials such as uranium, granite, and shale contain more concentrated amounts.

Radon that is released as it seeps up through the soil does not pose health risks. However, if radon seeps through cracks in a basement from rocks and sediment below that contain concentrations of radon, that house could be at risk of indoor air pollution. Once inside the home, radon can build to high levels, especially in unventilated areas.

After smoking, long-term exposure to radon is the second leading cause of lung cancer. In 1998, the EPA and U.S. Surgeon General's Office recommended that all people living in detached homes, town houses, mobile homes, or the first three floors of apartment buildings test their homes for radon levels, which should be monitored in the main living areas. If levels are found to be high, mitigation may include sealing cracks in the foundation and walls, increasing ventilation, and using fans to create more air circulation in these areas.

WHAT TO DO ABOUT AIR POLLUTION

The picture painted in the previous sections seems bleak; air pollution is a problem both inside and outdoors, and seems inescapable. In the past, the approach to solving the pollution problem was to disperse pollutants high into the atmosphere, away from the source, often through smokestacks. Today we now know that this is not a viable solution because dispersed pollutants are one of the biggest problems we face. New regulations aimed at reducing pollution and clean air legislations have begun to improve air quality.

REDUCING POLLUTION

The best approach to controlling pollution is to reduce activities that cause pollution in the first place. Recall that most air pollution is caused by the processes of transportation and energy production. Reducing electricity consumption, better insulation of homes, and improvements in public transportation are important first steps to reducing air pollution.

Technological advances are also critical to reducing harmful air pollution. There are many new devices aimed at reducing or controlling some of the major types of pollution we discussed. For example, electrostatic precipitators are used to control particulate matter in power plants. Though these devices require large amounts of electricity, they can collect as much as 99 percent of the particulate matter. It is important to note that the ash collected by this technique is solid waste that is often hazardous (containing heavy metals, for example) and must be disposed of properly.

Reducing the amount of soft coal that is burned will in turn reduce the amount of sulfur entering the atmosphere. This is important, as you will recall, because sulfur oxides are some of the chemicals most hazardous to human health. Changing from soft coal to natural gas reduces the amount of sulfur entering the atmosphere. However, this change is not always possible because of limitations of natural resources. Coal can be "cleaned" by crushing, washing, and then gassing it to remove the sulfur and methane before it is burned. However, the question then arises: Are we substituting a solid waste problem for a cleaner atmosphere, because the sulfur and metals will need to be disposed of?

Another innovation that has helped tremendously in the reduction of air pollution is the improvement of the internal combustion engine, greatly reducing the amount of nitrogen oxides that enter the atmosphere. For example, a car's catalytic converter typically uses catalysts that can remove up to 90 percent of the hydrocarbon, nitrogen oxide, and carbon monoxide emissions. However, despite these innovations, mobile sources such as cars are responsible for nearly 50 percent of cancers that result from outdoor air pollution, so there is still a long way to go.

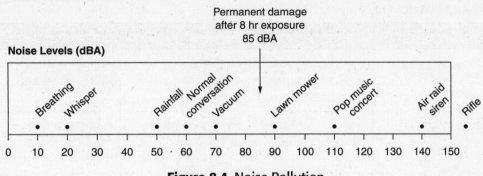

Figure 8.4 Noise Pollution

LEGISLATION

The Clean Air Act passed in 1963 was the first national legislation directed toward controlling air pollution when it had become clear that pollution was not a local problem. The Clean Air Act was heavily amended in 1970 with two goals in mind. First, a set of primary standards was established to protect human health. A set of secondary standards was also passed to protect materials, climate, crops, visibility, and personal comfort.

The Clean Air Act has been modified and updated since then. It now covers issues such as improved standards for sport utility vehicles and personal water crafts, and ozone protection, by phasing out CFCs, acid rain, and urban smog.

NOISE POLLUTION

Think about how noise, odor, and light pollution can reduce the quality of your life. Though these nuisances are not usually life-threatening, they can lead to the degradation of your environment. Noise pollution is often a problem in urban areas, and it is categorized as any unwanted, harmful, or disturbing sound that interferes with hearing, causes stress, disrupts concentration, or causes accidents. Permanent hearing damage in humans will begin after eight hours of exposure to noises at or above 85 dbA. (See Figure 8.4.)

WATER POLLUTION

Any chemical, biological, or physical change in water quality that has a harmful impact on living organisms is water pollution. Water pollution can be split into two main categories: pollutants that can cause health problems and pollutants that disrupt the ecosystem. These have been outlined in Table 8.1.

There are natural causes of water pollution, as shown in Table 8.1. These include sediment from erosion and natural oil seeps.

Table 8.1 Water Pollutants

Category	Examples	Dangers	Sources
Infectious agents	Bacteria, viruses, parasitic worms, and protozoa	Diseases	Animal wastes, untreated sewage
Wastes demanding oxygen	Organic wastes (animal manure)	Dying populations of fish	Sewage, animal feed lots, paper mills
Inorganic chemicals	Acids, lead, arsenic, selenium, salts	Unsafe drinking water, skin cancer, liver and kidney damage, harm to aquatic life	Surface runoff, household cleaners
Organic chemicals	Oil, gasoline, plastic, pesticides, detergents	Nervous system damage, cancer, harmful to fish and wildlife	Surface runoff, household cleaners
Plant nutrients	Nitrates, phosphates, and ammonium	Excessive algal growth, dangers to unborn babies and children, fish kills	Sewage, manure, runoff
Sediment	Soil, salt	Reduces photosynthesis, disrupts food webs, carries pesticides	Land erosion
Heat (thermal)	Excessive heat	Lowers oxygen level, thermal shock of organisms	Water cooling of electric and nuclear plants

In order to effectively identify and consequently clean up, control, or monitor water pollution, the source of the pollution must first be identified. Pollution control regulations are usually aimed at identifying whether the pollution is coming from a point source or a nonpoint source. **Point sources** discharge pollution from specific, identifiable locations. Factories, power plants, sewage treatment plants, coal mines, or oil wells can also be classified as point sources of pollution. As you can imagine, these sources are relatively easy to monitor and regulate. Quite often, in fact, it is possible to divert or treat the pollution before it enters into the environment.

Nonpoint sources of pollution are much more difficult to regulate and monitor than point sources because they have no specific points of origin. Pollutants such as runoff from agricultural fields, lawns, construction sites, or parking lots are all examples of nonpoint source pollution. Heavy rains

Point source

Nonpoint source

Figure 8.5 Point and Nonpoint Sources of Pollution

or snowmelt may spread concentrations of gasoline, lead, oil, or acid precipitation throughout the environment. (See Figure 8.5.) Runoff from terrestrial sources is the single biggest source of ocean pollution.

Contaminants carried by air currents can be considered a nonpoint source of water pollution. These contaminants are carried by air currents and then fall to the ground as various types of precipitation. One example of where this has happened is the Great Lakes. Tests have found large amounts of PCBs and dioxins as well as agricultural toxins in the Great Lakes that are too high to have come solely from local sources. In many cases, the pollutants' sources are thought to be thousands of kilometers away. The existence of these pollutants in the Great Lakes is possible only if they were carried through the atmosphere and then precipitated into the lakes. Because the pollutants were carried such great distances, it would be nearly impossible to regulate or monitor their movement.

GROUNDWATER POLLUTION

Nearly 50 percent of people in the United States depend on underground aquifers for their drinking water. Many people assume that natural processes within aquifers clean any contaminated water. In many cases, this is true; water moving slowly through the permeable layers of an aquifer is cleaned by this process. However, overuse, along with domestic, agricultural, and industrial pollutants, has begun to stress these reservoirs.

Methyl tertiary butyl ether (MTBE) is a gasoline additive that has been used since the 1970s to reduce the amount of carbon monoxide and ozone in automobile exhaust. MTBE is also believed to be a carcinogen and is one of the leading sources of groundwater pollution in the United States. MTBE is seeping into aquifers mainly from underground fuel storage tanks.

Fertilizers and pesticides are common pollutants of groundwater in farming communities. Herbicides end up in drinking water. Nitrates, found in fertilizers, often exceed safety standards in rural water supplies and can lead to health problems in infants. In Florida, 1,000 drinking wells were shut down because of excessive amounts of toxic chemicals from pesticides. Chemicals and wastes from agriculture are the largest sources of groundwater pollution. The **residence time**, or amount of time the water stays in an aquifer, can be very long. This means that some contaminants are extremely stable underground. Though it is possible to pump the water from the aquifer, clean it, and then pump it back in, that is a very expensive process.

SEWAGE TREATMENT

Human and animal wastes typically create the most serious health-related water pollution problems. More than 500 types of disease-causing bacteria, viruses, and parasites can travel from human or animal wastes into a given water supply. Biological agents have different levels of **pathogenicity**, which means they differ in the severity and virulence of illnesses they cause. Many of the most virulent agents do not spread through water at all, but are airborne. The EPA sets limits on the acceptable concentration of common biological agents in water supplies. For instance, for a water body to be considered safe for swimming, fecal coliform bacteria can have a concentration of no higher than 200 colonies/mL of water. How can the spread of water-borne diseases be prevented? A number of ways are discussed in the following sections.

SEPTIC SYSTEMS

In areas with small populations, natural processes can handle the buildup of human and animal wastes. As populations grow and become more concentrated, the natural systems cannot handle the accumulating wastes, diseases are spread, and living conditions are generally unpleasant.

Many cultures use human waste and animal manure as fertilizer for their crops. Although this provides valuable nutrients for the crops, it also introduces disease-causing pathogens into the food supply.

Septic tanks were developed to improve public health in rural areas. A typical septic system has water draining into a septic tank. (See Figure 8.6.) Substances such as grease and oil rise to the top while the solids sink to the bottom. Here the solids may decompose. The resulting liquid is channeled out through drain pipes that are surrounded by gravel. Most of the pathogens are killed as they are exposed to the oxygen in the soil. The excess water percolates through the gravel and eventually evaporates. Every so often, the effluent in the tank must be pumped out and taken to a treatment plant.

Individual or smaller septic systems work well in rural areas or smaller communities. But urban sprawl and city growth has demanded a solution adequate for a higher number of people.

MUNICIPAL SEWAGE TREATMENT

There are typically three steps in treating municipal solid waste. (See Figure 8.7.) The first is **primary treatment**. In this stage, raw sewage enters a treatment plant and the large solids are separated from the general wastes. A series of screens removes large debris and smaller objects.

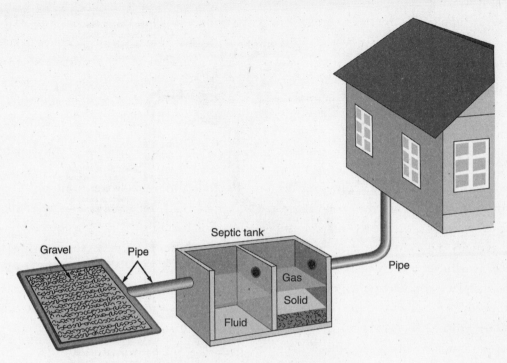

Figure 8.6 Domestic Septic System

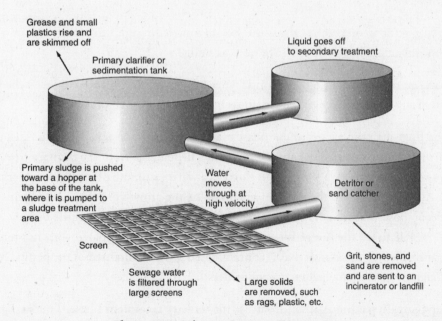

Figure 8.7 Primary Sewage Treatment

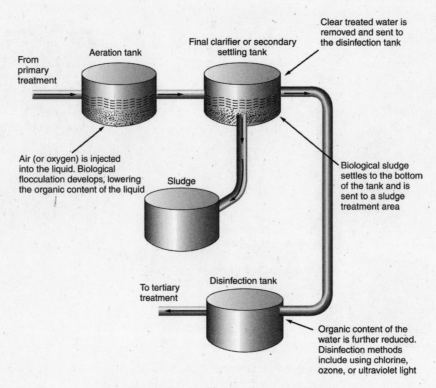

Figure 8.8 Secondary Sewage Treatment

Sand and gravel settle in one tank as the waste stream moves through a different tank in which about half the organic solids settle to the bottom as sludge.

The waste stream then enters into the **secondary treatment** phase. This step involves the biological degradation of any dissolved organic compounds in the water. (See Figure 8.8.)

The material from the primary treatment phase may flow through an aeration tank. This process is sometimes referred to as **activated sludge process**. The material from the primary treatment phase mixes with a bacteria-rich slurry. This slurry is a watery mixture of insoluble material as well as bacteria. Air is pumped into the mixture, encouraging the growth of bacteria that helps organic matter to decompose. The result is water in the top of the tank and sludge in the bottom. It is sometimes possible to use the sludge for fertilizer (if it was not contaminated with substances like heavy metals). However, the sludge that is contaminated must be disposed of properly. This often leads to a major cost for municipal treatment centers.

The last step in treating municipal solid waste is the **tertiary treatment** phase. This step removes plant nutrients, in particular nitrates and phosphates, from the end result of the secondary phase. (See Figure 8.9.)

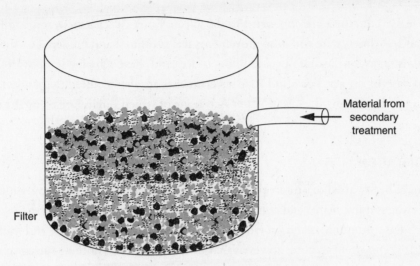

Figure 8.9 Tertiary Sewage Treatment

Treated wastewater is typically free of pathogens or organic matter and may contain high levels of inorganic nutrients. If released into the surface waters, this may lead to algal blooms or eutrophication. **Eutrophication** is the term given to the natural nutrient enrichment of a lake. This occurs mainly from the runoff of plant nutrients such as nitrates and phosphates. It is possible that during hot weather or times of drought, this overload of nutrients can lead to a dense growth of algae and cyanobacteria. Such conditions are called algal blooms. Passing the wastewater through a lagoon or wetland may reduce the effects of eutrophication and an algal bloom. The wastewater can also be added to a filter where chemicals or bacteria can be used to bind the nutrients in question.

CLEAN WATER ACT

The goal of the Clean Water Act, passed by the U.S. Congress in 1972, was to return all of the surface waters in the United States to "fishable and swimmable" conditions. With this act, legislators brought water quality to the forefront of the people's minds. Maintaining and providing clean water for all citizens became a national priority.

The Clean Water Act requires that specific point sources of pollution acquire a permit and develop technology that would enable them to control their output. The Clean Water Act also requires the use of **best available, economically achievable technology** for limiting discharge of toxic substances and allows no discharge at all of 126 toxic pollutants.

As a result of these efforts, the surface water in the United States has improved dramatically. All the surface water is not yet swimmable or fishable, but the improvements have been significant. One of the most significant changes has been the allocation of federal and state money to build municipal sewage treatment facilities.

As with any legislation, there are opponents to the Clean Water Act. Farmers, some industries, developers, and, at times, state and local movements feel pressured and hindered by the Clean Water Act. Provisions for the draining and filling of wetlands have remained one of the more controversial parts of the act. Local and state governments are also required to spend money implementing and enforcing the Clean Water Act, none of which is reimbursed by the federal government.

More Legislation

Other laws have been passed to protect the waters in the United States. The Safe Drinking Water Act regulates water in municipal and commercial systems. Some people say that the regulations are too loose for rural communities and provide evidence of pesticides, herbicides, and lead to back up their claims. It is important to note, however, that the mere presence of these substances is not the same as having them at dangerous levels.

An important international agreement concerning water quality was signed in 1972. The Great Lakes Water Quality Agreement was signed by Canada and the United States. This agreement has made huge strides in cleaning up this major water system. In 1990, The London Dumping Convention set regulations for phasing out ocean dumping of industrial wastes, effluent, and plastics by 1995. The results of this international convention still remain to be seen.

REVIEW QUESTIONS

1. Dust from construction sites, strip mines, or soil erosion is an example of which type of air pollution?

   (A) Primary pollutant

   (B) Secondary pollutant

   (C) Fugitive emissions

   (D) Natural pollutants

   (E) Conventional pollutant

2. Which of the following is an extremely hazardous gas produced by the incomplete burning of fossil fuels?

   (A) Ozone

   (B) Carbon dioxide

   (C) Lead

   (D) Nitrous oxide

   (E) Carbon monoxide

3. What is the chemical formula for ozone?

   (A) O

   (B) O_2

   (C) O_3

   (D) CO_2

   (E) CH_4

4. Which statement most accurately describes a temperature inversion?

   (A) A stable layer of warmer air over cooler air

   (B) An unstable layer of warm air over cooler air

   (C) A layer of cooler air over warmer air

   (D) A mixing of all the layers of the atmosphere

   (E) A layer of ozone concentrated in the stratosphere

5. Which of the following is an indoor air pollutant found in particle board, furniture, wallpaper, and carpeting?

   (A) Ozone

   (B) Nitrous oxide

   (C) Radon

   (D) Formaldehyde

   (E) MBTE

6. Which of the following kills pathogens and removes most organic material through aeration?

   (A) Septic systems

   (B) Primary sewage treatment

   (C) Secondary sewage treatment

   (D) Tertiary sewage treatment

   (E) Municipal septic systems

7. Runoff from construction sites is an example of which of the following?

   (A) Primary pollution

   (B) Nonpoint source pollution

   (C) Point source pollution

   (D) Heat island

   (E) Tertiary sewage

8. What is the natural enrichment of a lake environment called?

   (A) Primary treatment

   (B) Secondary treatment

   (C) Tertiary treatment

   (D) Eutrophication

   (E) Algal bloom

9. Phasing out the use of CFCs and reducing smog is covered under which of the following?

   (A) Clean Water Act
   (B) Safe Drinking Water Act
   (C) Clean Air Act
   (D) Fugitive Emissions Act
   (E) Environmental Protection Agency

10. Which of the following is a point source of water pollution?

    (A) Smoke stack
    (B) Outflow pipe
    (C) Runoff from streams
    (D) Snowmelt
    (E) Soil erosion

FREE-RESPONSE QUESTION

In 1969, the Cuyahoga River in Ohio caught fire, an embarrassing spectacle that led to a new era of environmental regulation in the United States. This was actually just one of several burning water bodies that raised public awareness about water pollution, while around the same time visibly terrible air quality led to similar pushes for curbs on air pollution.

(A) (2 points) What are some pollutants that might result in a water body catching fire, and what impact might these have on the aquatic environment?

(B) (2 points) Describe one possible point source for this pollution and one possible nonpoint source.

(C) (3 points) Name and describe the central tenets of one environmental law in the United States that deals with preventing or limiting water pollution.

(D) (3 points) You are the mayor of a small city that sits on the edge of a large lake. What steps can you take to limit the discharge of nonpoint source pollution into the lake?

_____

_____

_____

_____

_____

_____

_____

_____

_____

_____

_____

_____

_____

ANSWERS AND EXPLANATIONS

1. C

A primary pollutant is released directly from the source into the air while a secondary pollutant is modified by reactions in the air. This eliminates choices (A) and (B). Dust from a construction site or a strip mine is not a natural pollutant, (D), because the dust could have been caused by human activity. And a conventional pollutant, (E), is something like sulfur dioxide, carbon dioxide, or lead.

2. E

Ozone is formed when nitrogen oxide splits in the atmosphere, releasing an oxygen atom to join with free oxygen in the air, so choice (A) is incorrect. Lead is a solid, not a gas, so choice (C) is incorrect. Choice (B) is incorrect because carbon dioxide is also released during the burning of fossil fuels, but it is not extremely hazardous. Nitrous oxide is a reactive gas that forms when air is heated to about 650°C, so choice (D) is incorrect.

3. C

There are three molecules of oxygen in ozone, which eliminates choices (A) and (B) as possible answers. Choice (D) is incorrect because CO_2 is the formula for carbon dioxide, and choice (E) is incorrect because methane has the formula CH_4.

4. A

Choice (B) is incorrect because a temperature inversion involves a stable level of warm air, not an unstable one. Choice (C) describes a "normal" situation in the atmosphere, so this answer is incorrect. A temperature inversion prevents mixing of the atmosphere, which eliminates (D) as an answer. And because it describes the ozone layer, not a temperature inversion, choice (E) is incorrect.

5. D

Ozone and nitrous oxide are both outside air pollutants, so choices (A) and (B) are incorrect. Radon is a gas that is often found accumulated in the lower levels of buildings, so choice (C) is incorrect. Choice (E) is incorrect because MBTE is an additive to gasoline that can contaminate a water supply.

6. C

Septic systems are for residential use and are quite simple. This eliminates choices (A) and (E). Primary sewage treatment separates large solids from a waste stream, making (B) incorrect. And choice (D) removes plant nutrients.

7. B

Primary pollution, (A), and point source pollution, (C), are types of pollution that come directly from specific sources. Construction sites are dispersed and are not a source of one specific type of pollutant, eliminating these choices. Nonpoint source pollution, (B), refers to pollution from grouped or difficult-to-identify sources such as that from agricultural fields or, in this case, construction sites. Tertiary sewage, (E), is a phase in the treatment of solid waste. A heat island, (D), is not related to pollution in this manner.

8. D

Choices (A), (B), and (C) are all phases in the treatment of solid waste. An algal bloom, (E), may occur as a result of the natural enrichment of a lake, leaving (D) as the correct answer.

9. C

Choices (A) and (B) are legislation for clean water, not air. There is no Fugitive Emissions Act, (D), and the EPA, (E), is a government agency. This leaves choice (C) as the correct answer.

10. B

A smoke stack, (A), is a point source for air pollution, not water pollution. Runoff, (C), snowmelt, (D), and soil erosion, (E), are nonpoint sources of pollution.

ANSWER TO FREE-RESPONSE QUESTION

(A) There are many answers to this question, as many pollutants are flammable and light enough to float on water. Oil, gasoline, and almost any form of fuel are all possibilities, and oil in particular was implicated in the 1969 fire. There are many other floating flammable chemicals, such as acetone, turpentine, or benzene, but these evaporate rapidly, making fire or explosion possible but more unlikely. Another option is solid debris such as paper, polystyrene foam, or plastic; any of these materials can burn.

The impact on aquatic life depends on the pollutants you choose to discuss. Oil, for instance, can coat many aquatic organisms, resulting in suffocation and death. When oil coats aquatic plants, the result can be reduced levels of dissolved oxygen in the water, further stressing aquatic life. Oil may also remove waterproofing from aquatic birds and mammals, resulting in hypothermia.

(B) The answer to this question also depends on which pollutants you choose. Taking oil as an example, a point source can be any place that stores oil, such as a leaking refinery tank, rail depot, or even a crashed tanker truck. Nonpoint sources for oil might include parking lots, roads, or construction sites. Solid debris does not have a point source, although some solid debris from nonpoint sources may dissolve and leak chemicals into storm water. An example of a solid debris nonpoint source would be a poorly sealed municipal dump.

(C) There are many options to choose from, such as the Clean Water Act of 1972, which set standards for water quality, established regulations for discharges into water bodies, and mandated water treatment for municipal waste water. Another option is the Oil Pollution Act of 1990, which established new regulations for oil spill response capacity and assessment of damage to the environment. There are other laws that indirectly impact water quality, such as the Comprehensive Environmental Response and Liability Act of 1980, which established the Superfund for cleanup of contaminated sites, or the National Environmental Policy Act of 1969, which set the stage for almost all subsequent environmental regulations and mandated Environmental Impact Statements for anything the government does.

(D) There are many ways to answer this question, depending on which sources you choose to focus on, but some general ways to reduce nonpoint source pollution are to encourage buffer zones such as wetlands along the shoreline or along roads, and to reduce impermeable surfaces by encouraging open spaces such as parks. There are also regulatory approaches, such as passing local laws to discourage fertilizer overuse or mandating inspection of storage tanks and vehicles for leaks.

CHAPTER 9: ENVIRONMENTAL CHANGE AND THE FUTURE

IF YOU LEARN ONLY FIVE THINGS IN THIS CHAPTER . . .

1. Stratospheric ozone protects Earth from harmful ultraviolet radiation from the Sun. A thinning of the ozone layer can lead to a rise in temperature in the troposphere.

2. Greenhouse gases absorb heat and trap it close to the surface of Earth.

3. Global warming and global climate change are not the same thing, but they are related. Global warming can lead to an overall change in global climate.

4. Biodiversity is a natural resource that must be protected. Humans depend on biodiversity for food, medicines, fibers, aesthetic reasons, and to maintain the health of their water and air supplies.

5. Maintaining environmental health is a global issue. Developed and developing nations need to work together to institute scientific advances and regulations that ensure clean water and air, reduce emissions, and maintain biodiversity.

As the human population continues to grow exponentially and require more resources, there are questions about the world being left behind for future generations. Issues such as global warming and the loss of habitats and species are growing concerns.

THE STATE OF THE STRATOSPHERIC OZONE

How can ozone—an air pollutant that is a major component in photochemical smog—be such an important component in the upper atmosphere that there are grave concerns about its depletion?

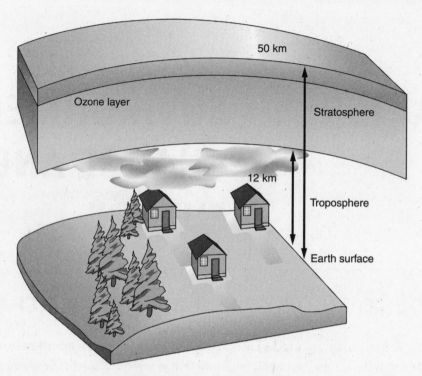

Figure 9.1 The Stratosphere

Ozone is found naturally in the stratosphere where it actually protects Earth from harmful radiation. It is only hazardous to health and property when found in the troposphere. The stratosphere is less dense than the troposphere below it, but it does basically have the same composition, except for two notable exceptions: There is much less water vapor in the stratosphere, and there is much, much more ozone. (See Figure 9.1.)

THE FORMATION OF OZONE

Ozone (O_3) forms when atomic oxygen reacts with molecules of O_2. Single atoms of oxygen are often the result of secondary atmospheric reactions that are driven by energy from the Sun. Ultraviolet light from the Sun contains the energy necessary to break apart molecules of oxygen. A molecule of oxygen in the stratosphere will absorb the ultraviolet radiation from the Sun and break apart, forming two atoms of oxygen. These oxygen atoms are now free to react with a molecule of oxygen (O_2) to form a molecule of ozone (O_3). This reaction actually proceeds faster at lower altitudes, but the reaction that splits O_2 into two atoms of oxygen is faster in the upper atmosphere. Much of the ozone found in the stratosphere is formed in the tropics, where the Sun is almost directly overhead for the entire year. The ozone is then distributed around Earth by winds in the stratosphere.

Lightning is also responsible for adding ozone to the upper atmosphere. Many lightning storms create nitrogen oxides. These nitrogen oxides then react with sunlight to produce ozone in the same way that ultraviolet light does. Because lightning occurs in conjunction with a storm, the ozone is not produced near the surface because of the lack of sunlight. Therefore, lightning does not increase the ozone at the surface, but studies have shown that storms in tropical regions can add ozone to the upper atmosphere. As stated earlier, ozone in the stratosphere protects Earth from the Sun's harmful ultraviolet rays (covered in the next section). Ozone in the lower atmosphere, however, is a different story. Here, O_3 is an oxidizing reagent and damages vegetation and building materials (e.g., paint, plastics, rubber) and can even damage tissues in the lungs and eyes.

WHY DO WE NEED THE OZONE LAYER?

The Sun emits powerful radiation in the form of the electromagnetic spectrum. (See Figure 9.2.) Radiation with a wavelength shorter than visible light contains enough energy to damage organic molecules, including proteins and nucleic acids. Ultraviolet radiation falls within that range, with wavelengths between about 10 nm to 0.1 μm. It is extremely harmful to living things and has been directly linked to skin cancer, genetic mutations, crop failures, and disruption of biological communities.

UVA is the least energetic form of the ultraviolet radiation, with a wavelength between 315–400 nm, and is not absorbed by the ozone layer. UVB, a form of ultraviolet radiation with a wavelength between 290–330 nm, on the other hand, is absorbed, as is UVC. UVB radiation is associated with several health problems, including cataracts and basal cell carcinoma, the most common form of skin cancer. With wavelengths of 100–280 nm, UVC is the most energetic ultraviolet light, and therefore potentially most damaging. For now, we are protected from most UVC radiation, but if the ozone layer were to continue to break down, that protection would be jeopardized.

OZONE DEPLETION

In 1985, the British Antarctic Atmospheric Survey made a startling discovery when they found that the stratospheric ozone levels over the South Pole had dropped significantly during the months

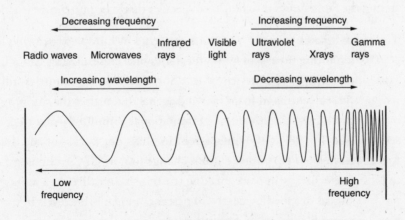

Electromagnetic Spectrum

Figure 9.2 The Electromagnetic Spectrum

of September and October. It is believed that this thinning had been occurring since at least the 1960s, but the miscalibration of instruments led to incorrect reporting.

In 1998, an estimated 10 percent of all the stratospheric ozone worldwide was destroyed; at that time the Arctic average was 40 percent below normal. In 2000, the thinned section of the ozone layer above Antarctica was about 44.3 million square kilometers. Estimates indicate that a loss of 1 percent of the ozone from the stratosphere leads to a 2 percent increase in the amount of ultraviolet radiation reaching Earth's surface. This could lead to as many as a million extra cases of skin cancer each year. These alarming facts are evidence that this problem is real and needs to be addressed.

WHY IS THIS HAPPENING?

Chlorine is the main culprit in human contribution to the depletion of the ozone layer. **Chlorofluorocarbons** (CFCs) are chlorine-containing molecules that we release into the atmosphere. CFC, sometimes known as Freon, was invented in 1928 and was considered nontoxic, nonflammable, chemically inert, cheap, and useful in many applications. Items such as refrigerators, air conditioners, and aerosol cans used CFCs. Because the atoms in CFCs are so stable, they remain in the environment for a long time. As they pass into the stratosphere, ultraviolet radiation releases the chlorine atoms. Because the chlorine atoms are not consumed in the reaction, they can remain in the stratosphere for a long time, thereby destroying ozone. Even more effective at depleting ozone is bromine, chemically similar to chlorine and used extensively in fire suppressants. This gas is also released naturally in trace amounts by some phytoplankton.

Temperatures in the stratosphere above Antarctica drop so low in that area that ice crystals form, a phenomenon that rarely happens anywhere else on Earth. Molecules that contain chlorine are absorbed into the surfaces of the ice crystals. During the spring, the Sun releases the chlorine molecules that speed up the destruction of ozone.

The United States, Canada, and some countries in Europe passed laws in the late 1970s prohibiting the use of CFCs for nonessential uses. Using CFCs in aerosol cans is one example of nonessential use. Some developing countries still use CFCs as refrigerants, solvents, and foam blowing agents. Soon after the thinning ozone layer over Antarctica was discovered, the international community met to deal with this serious issue. In 1987, 81 nations agreed to phase out the use of all CFCs. A fund was established to help poorer nations change to non-CFC technology. Alternatives to CFCs were already in place, making the transition easier. As a result, CFC production has dropped by nearly 80 percent, and chlorine levels in the atmosphere have been steadily declining. Scientific models predict that the levels of ozone in the stratosphere could return to normal by 2040.

AP EXPERT TIP

After reading about the formation and breakdown of ozone, find a friend or family member and describe the steps involved in each. Without using the book, "teach" the person about the formation and breakdown of ozone. This will help you remember the info on Test Day.

GLOBAL WARMING: CAUSES AND EFFECTS

Global warming is a rise in temperature in the troposphere that may lead to climate change. **Global climate change** is a much broader term that discusses changes in any part of the climate on Earth. These changes could occur in temperature, precipitation, or storm intensity. A change in global climate could be a warming or a cooling of the temperatures.

Often coupled with a discussion of global warming is talk of the **Greenhouse Effect**. The **greenhouse gases**—such as water vapor, carbon dioxide, methane, and nitrous oxides—and particulate matter in the atmosphere reflect about 30 percent of radiation from the Sun back into space. The rest of the radiation that makes it to Earth's surface is then either reflected back into space by surfaces such as snow and sand, or it is absorbed. Once absorbed, it can be re-emitted into the atmosphere as **thermal infrared radiation**, but this radiation has a longer wavelength. (See Figure 3.14 on page 63.)

Human activity, however, can release more greenhouse gases into the atmosphere, raising the level of these gases to a higher level than normal. The increase in carbon dioxide is the most significant impact that humans have had on the addition of greenhouse gases to the atmosphere. The burning of fossil fuels and forests among other human activities releases nearly 30 billion tons of CO_2 each year. Methane, or CH_4, absorbs more infrared radiation and is accumulating in the atmosphere faster than CO_2. CH_4 is released by animals such as cows and bison, and from rice fields, coal mines, and landfills (the largest source). Burning organic material releases nitric oxide, NO_2. The higher level of gases can enhance the natural greenhouse and lead to global warming.

THE KEELING CURVE AND THE "HOCKEY STICK"

How do scientists know that CO_2 is linked to climate change? There have been many investigations, but two pioneering studies set the stage. The Keeling Curve, a graph based on atmospheric CO_2 readings at Mauna Loa Observatory in Hawaii and named for its discoverer, Charles Keeling, showed just how much CO_2 human activity was adding to the atmosphere over the past 50 years. The second, the "Hockey Stick," was developed by many researchers and shows global temperature records going back thousands of years. It shows a relatively flat graph until the 18th century (the start of the Industrial Revolution), at which point global temperature climbs rapidly, giving the graph the general shape of a hockey stick. The "Hockey Stick" is based on data from ice cores, tree rings, isotopes, and other sources.

AP EXPERT TIP

Some greenhouse gases have an impact unrelated to global warming. For example, CO_2 is a weak acid when dissolved in seawater, and high concentrations can lower the pH of the oceans. Lowered pH in turn dissolves the calcite shells of corals and other marine organisms, killing them, which results in severe disruption to marine ecosystems.

IMPACTS OF GLOBAL WARMING

Temperatures in the troposphere are strongly influenced by the snow and ice near the North and South Poles and in mountain glaciers. Global air circulation causes the temperature to increase more at the poles as excess heat from the equator is transferred toward them. Information about global air circulation has allowed scientists to keep close watch on what is happening at the poles and project what this means for the climate of the entire planet. Measurements now show floating ice in the seas around the North Pole and a thinning ice sheet in Greenland. In addition, over the past 60 years (since 1950), the average temperature of the upper 300 meters of the ocean, also known as the sea surface, has increased by around 0.31°C, and the average temperature of the upper 3,000 meters has increased by around 0.06°C. This may not sound like a lot, but during the period from 1870 to 1910, only 100 to 140 years ago, the average sea surface temperature did not change at all.

Less ice in the Arctic has potentially serious consequences. Areas of ice and snow help cool the Earth by reflecting 80 percent to 90 percent of the sunlight that makes it to the surface back into space. More sunlight is absorbed by darker regions such as cities, forests, grass, and oceans. A rise in the troposphere's temperatures can cause some of Earth's ice to melt. As the abundance of ice is reduced, temperatures in the troposphere continue to rise.

WARMER TEMPERATURES: GOOD OR BAD?

Perhaps the greatest negative impact of global warming is ecosystem disruption and biodiversity loss, as species and ecosystems are unable to adapt to rising temperatures. Some ecosystems, such as coral reefs, are particularly vulnerable. Warming temperatures are also expected to facilitate invasions by exotic species, with further negative consequences for biodiversity. There are several other potential negative consequences, including accelerated conversion of arable land into desert, particularly in sub-Saharan Africa. There is even a potential public health risk, as tropical diseases expand their range in a warmer world. To put everything in perspective, during the Last Glacial Maximum (Ice Age), average global temperatures were only 5°–9°C (10°–19°F) lower than today. A change of a few degrees can fundamentally alter ecosystems across the planet.

Another serious, negative consequence of global warming is the melting of the polar ice caps in Greenland and Antarctica. Sea level could rise by as much as 100 meters as a result. This is especially alarming considering that one-third of the world's population currently lives in areas that would be flooded were this to occur. Many mountain tops may lose their glaciers, causing great trouble for areas that depend on freshwater from the partially melting glaciers for reservoirs and farming.

As global weather patterns shift due to a rise in the troposphere's temperature, rains could come more frequently to higher latitudes, increasing crop production in those areas, although this might be offset by increased risk of drought at lower latitudes. Increased amounts of CO_2 in the atmosphere could benefit plant growth because CO_2 serves as a fertilizer. This could increase crop production in some areas, providing more food and boosting agriculturally based economies.

Warmer temperatures are also melting the **permafrost** beneath the arctic soils in parts of Alaska. As this happens, CO_2 and CH_4 are released, further increasing the rate of warming. Some of these greenhouse gases can be reabsorbed by natural habitats, particularly forests and wetlands, but as these areas are lost to development, less and less carbon may be taken up this way. As a result, there should be an extended growing period in colder climates, increased tourism, and ports that are no longer clogged with ice. However, buildings, roads, and parts of the Trans-Alaska Pipeline are sinking, shifting, or even breaking apart.

HOW CAN WE REDUCE OR CONTROL CLIMATE CHANGE?

The international community has begun to focus its efforts on reducing and stabilizing greenhouse gas emissions with the goal of reducing the threat of global warming. The **Kyoto Protocol** was signed in 1997 by 160 nations who agreed to reduce carbon dioxide, methane, and nitrous oxide emissions to about 5 percent below 1990 levels by 2012. It was also agreed that three other greenhouse gases—fluorocarbons, perfluorocarbons, and sulfur hexafluoride—would be reduced as well, although the levels were not determined.

Progress has been made toward reducing greenhouse gas emissions. Great Britain, for example, reached the 1990 CO_2 emission levels by the year 2000. They have started to substitute natural gas for coal, have increased energy efficiency in homes and businesses, and have raised their gasoline tax. Many countries are on the right track, and levels of other greenhouse gases are coming down as well. The United States signed but has not ratified the Kyoto Protocol. This means that the United States, which produces 28 percent of the human-made CO_2, could be 43 percent above 1990 emission levels by 2020.

A LOSS OF BIODIVERSITY

Biological diversity, or **biodiversity**, is one of the most important natural resources on Earth. Biodiversity supplies food, wood, fibers, energy, medicine, and more. It also helps maintain the quality of our air and water, keeps the soils fertile, and controls pest populations.

Concepts used when talking about biodiversity include the following:

- **Genetic diversity**—a variety of genetic material within a species or a population
- **Species diversity**—the number of species in different habitats
- **Ecological diversity**—variety of terrestrial and aquatic ecosystems
- **Functional diversity**—biological and chemical processes needed for the survival of ecosystems

Human threats to biodiversity can be summarized by the acronym HIPPO: **H**abitat destruction, **I**nvasive species, **P**ollution, **P**opulation, and **O**verharvesting.

Often, the greatest threat to species and biodiversity is the loss, deterioration, or fragmentation of habitats. Habitat loss has the greatest impact on organisms. Clear-cutting forests has destroyed the habitat of countless species. Forests today cover less than half the areas they once did, and many grasslands have been converted to farms or housing. Habitats have also been destroyed by the side effects of mining, dam building, and various fishing methods.

Some small, scattered areas of habitat are maintained, but this is not always enough to preserve all species in the area. Large mammals, such as tigers or wolves, need considerable expanses of undisturbed land. If a species' habitat becomes fragmented or disrupted, the species becomes more susceptible to severe weather, disease, and genetic flaws as members of their community die off and inbreeding increases.

Invasive species, or alien species, are also a major threat to native biodiversity. Their new homes greet them with a lack of predators, disease, and more resources than their old environment provided. Invasive species in North America include the kudzu vine, purple loosestrife, gypsy moth, mongoose, sea lamprey, zebra mussel, and nutria. The rate of influx of invasive species has increased significantly in recent years due to ease and popularity of travel by air, water, and land. People may bring a plant or animal into the country deliberately, or the species may hitch a ride on a boat hull, in a house plant, in packing crates, or even in the dirt on the bottom of a shoe. As covered in Chapter 6, overharvesting activities such as overfishing are a new reality because commercial fishing has become so technical and efficient; fleets are able to make large catches on single trips. In Chapter 8, we saw the complex interactions between pollution and biodiversity.

The biggest concern about the loss of biodiversity is, of course, the threat of **extinction**. Extinction is the complete elimination of a species. It is important to realize, however, that extinction is a natural process. Even in undisturbed ecosystems, about one species goes extinct every 10 years. In the past 100 years or so, however, human impact on ecosystems has accelerated this process dramatically. If we continue on our current track, millions of plants and animals and microbes may disappear from Earth in the next few decades.

LAWS AND PROGRAMS TO PROTECT BIODIVERSITY

Nations around the world are working to reduce the HIPPO threats and maintain the biodiversity that now exists. The United States has enacted many laws and programs to safeguard the environment for future generations. The most well known, perhaps, was the establishment of the Environmental Protection Agency (EPA) by President Nixon in 1970. Its function is to protect human health as well as protect and preserve Earth's air, water, land, and endangered species. The Clean Air and Water Acts, as well as the Superfund for toxic waste disposal, are other examples of U.S. legislations. Launched under the administration of George H. W. Bush in 1992, Energy Star is a program to promote energy-efficient consumer products. Participating companies can indicate their involvement in the program by using the official Energy Star label. In 2006, under the administration of George W. Bush, the EPA launched WaterSense, whose goal is to foster water efficiency in the same vein as the Energy Star program.

THE ENDANGERED SPECIES ACT

A bill was written in 1874 to protect the American bison. The U.S. Congress failed to pass the bill because of the prevailing feelings of the day. Most people thought of wild animals as abundant, but this bill showed an early effort to protect biodiversity. By the turn of the 20th century, most states had some hunting and fishing restrictions in place. The thought was to preserve the wildlife for future human use, not to protect the species themselves. It was not until 1973, with the passage of the U.S. Endangered Species Act, that new approaches to wildlife protection were accepted. It sought to identify all endangered species and protect biodiversity, regardless of how useful a particular species is to humans.

An **endangered species** is one that is in imminent danger of extinction. A **threatened species** is one that is likely to become endangered in at least one local area. In some areas, bald eagles, brown bears, and sea otters are threatened species although their populations are normal in other areas. A **vulnerable species** is naturally rare or has been depleted by human activity and is unable to recover. The Endangered Species Act regulates activities that involve endangered species, such as harassing, harming, hunting, shooting, trapping, collecting, importing or exporting, possessing, or selling endangered species. Regulations under this act control live organisms, body parts, or products made from endangered species.

Many private groups and Non-Governmental Organizations (NGOs) also have programs in place that aim to preserve biodiversity. The United Nations, for instance, has several branches devoted to global environmental threats, including the United Nations Environment Program (UNEP) and United Nations Educational, Scientific, and Cultural Organization (UNESCO). The latter is also concerned with cultural heritage. Both programs, along with several international treaties, work to protect international biodiversity and cultural heritage. They use every means possible except direct management of sites or reserves, which is left up to individual nations.

Approaches range from buying land outright for protection, to lobbying national or local governments for better species protection laws, to helping build ecotourism projects in environmentally sensitive areas.

For extremely threatened species, **captive breeding** is an option. Highly endangered species are removed from the wild to be bred in zoos or other facilities, and one day may be returned to the wild when conditions improve. There have been some success stories, such as the black-footed ferret in the western United States, but the process is expensive and faces many hurdles, such as the difficulty of determining gender in animals such as fish.

ECONOMIC, SOCIAL, AND LEGAL OUTCOMES OF ENVIRONMENTAL CHANGE

Changing the state of the environment is a difficult and sometimes daunting task. There are ramifications on all of our lives as these changes are implemented. There are, however, often overlooked global implications. Protecting and managing the environment is an effort that has to come from all of us.

GLOBAL ECONOMIES

Globalization is the newly evolved interdependence of national economies that resulted from advances in communications, transportation, finances, and commerce has introduced opportunities for change as well as challenges to environmental management. International cooperation is needed to maintain a healthy environment. Questions about how environmental decisions are made and who makes them are all part of a process called **environmental governance**.

THE WORLD BANK

Third-world nations often have difficulty adhering to environmental policies established by developed nations. Quite often the policies are simply too cost prohibitive for these poorer nations. The World Bank was founded in 1945 to provide funds for Japan and countries in Europe to help them rebuild after World War II. In the 1950s, the World Bank shifted its focus and began sending aid to third-world nations.

Some projects funded by the World Bank have been highly controversial and have negatively impacted the environment. For example, the World Bank is funding construction of nearly 200 dams along a river in India. The dams are being constructed to provide hydroelectric power and water supplies to areas downstream of the river. However, the project will result in the flooding of upstream areas and the displacement of as many as 1.5 million people.

In response to incidents such as this one in India, the U.S. Congress now insists that all loans for international development be reviewed for potential environmental and social impacts. Each project must use renewable resources, must not cause severe or irreversible damage to the environment, and must not displace people.

TRAGEDY OF THE COMMONS

The **tragedy of the commons** refers to a conflict between individual interests and the common good with regard to the use of resources. In 1968, Garret Hardin argued in an article with this title that any commonly held resource would eventually and inevitably be degraded or destroyed, as people's self-interest outweighed the interests of the general public. While bleak, this scenario is only sometimes accurate. Pelagic fisheries such as the bluefin tuna fishery are classic examples of how common management has failed; however, many communal resources have been successfully managed and maintained. Examples include Native American management of hunting grounds, pastures and forests owned by villagers in Switzerland, Maine lobster fisheries, irrigation systems in Bali and Laos, and many near-shore fisheries. These **communal resource management systems** share the following features:

- Community members' long history on the land or near the resource anticipating this history to continue for generations
- Clearly defined resource
- Known community size
- Resource of relative scarcity, meaning that the community has to work together
- Resource closely monitored, discouraging overuse or cheating
- Conflict resolution methods in place when and if conflicts arise
- Incentives to encourage compliance with the rules

REVIEW QUESTIONS

1. Which of the following is NOT a greenhouse gas?

    (A) Water vapor
    (B) Carbon dioxide
    (C) Oxygen
    (D) Methane
    (E) Chlorofluorocarbons

2. Which of the following statements is most accurate?

    (A) Ozone near the surface of Earth protects us from ultraviolet radiation.
    (B) Ozone in the stratosphere protects us from harmful radiation from the Sun.
    (C) Exposure to ozone increases one's risk of skin cancer.
    (D) Ozone is added to the atmosphere by burning fossil fuels.
    (E) Ozone can form in the deep ocean.

3. What did the Kyoto Protocol propose to regulate?

    (A) The emissions of greenhouse gases
    (B) Ozone depletion
    (C) Loss of biological habitats
    (D) Aid to developing nations
    (E) Use of natural resources

4. Which term describes a species in imminent danger of disappearing forever?

    (A) Extinct
    (B) Endangered
    (C) Threatened
    (D) Invasive
    (E) Exotic

5. What role do chlorofluorocarbons play in the environment?

    (A) They add ozone to the troposphere.
    (B) They add ozone to the stratosphere.
    (C) They lead to the depletion of strato-spheric ozone.
    (D) They lead to the depletion of ground-level ozone.
    (E) They no longer pose a threat to the environment.

6. Why will melting the ice sheets and glaciers on Earth raise temperatures even more?

    (A) The melt water will raise global sea levels.
    (B) Snow reflects more heat than bare ground so more heat will be absorbed.
    (C) Snow and ice absorb heat that leads to higher temperatures.
    (D) The air will warm because there is less snow and ice.
    (E) Climates will change.

7. What is the greatest threat to terrestrial species?

    (A) Introduced species
    (B) Climate change
    (C) Pollution
    (D) Overharvesting
    (E) Habitat destruction

8. A species that is introduced to a new environment and thrives is called a(n)

    (A) endangered species.
    (B) extinct species.
    (C) threatened species.
    (D) invasive species.
    (E) keystone species.

9. Which statement BEST summarizes the Tragedy of the Commons?

   (A) Self-interests outweigh public interests.

   (B) Public interests outweigh self-interests.

   (C) It is all right to be selfish as long as you think of the big picture.

   (D) It is a tragedy when biodiversity disappears.

   (E) It is the responsibility of developed nations to help developing nations.

10. Which of the following provides development aid to developing nations?

   (A) The U.S. Congress

   (B) The Kyoto Protocol

   (C) The World Bank

   (D) The Environmental Protection Agency

   (E) The Endangered Species Act

FREE-RESPONSE QUESTION

A federal highway is being built in an area that is in the process of establishing a wildlife corridor. Guidelines must be set by the conservation biologists and government agencies working on the project.

(A) (2 points) Explain the meaning and function of wildlife corridors.

(B) (2 points) Name and describe one law that will be relevant in building this highway.

(C) (2 points) Name one animal that could be helped by this corridor and explain how the corridor will be used.

(D) (3 points) Describe two physical attributes or behavioral characteristics of a species that make it much more susceptible to becoming endangered or extinct.

_____

_____

_____

_____

_____

_____

_____

_____

_____

_____

_____

_____

_____

_____

_____

_____

ANSWERS AND EXPLANATIONS

1. C

Oxygen is present in the atmosphere but is not a greenhouse gas. The other four gases are greenhouse gases.

2. B

Ozone near the surface is very harmful to animals and plants, so this eliminates choice (A). Exposure to ozone, (C), does not increase Sun cancer risks, although a decrease in the ozone layer will lead to more ultraviolet radiation on Earth that can increase your chances of cancer. Ozone is added to the atmosphere when nitrous oxides are broken down, eliminating choice (D). Ozone needs sunlight to form, eliminating choice (E).

3. A

The Kyoto Protocol called for a reduction of the use of greenhouse gases to 1990 levels. This may lead to a reduction in the ozone "hole," but that was not its direct mission.

4. B

If a species is extinct, (A), it is gone forever. Endangered species, (B), are in danger of becoming extinct. Threatened species, (C), are reaching a level where they may become endangered. Invasive, (D), and exotic, (E), species are the same thing. They are non-native species that have been introduced to a new ecosystem.

5. C

Chlorofluorocarbons break down in the stratosphere, releasing chlorine atoms that speed up the destruction of ozone. Because they break down in the upper atmosphere, they do not deplete ground-level ozone, so (D) is incorrect. They do not add ozone to any part of the atmosphere, eliminating (A) and (B). Because the chlorine atoms are not destroyed in the upper atmosphere, they remain in the environment for a very long time, making (E) incorrect.

6. B

Snow does reflect more heat and sunlight, so without the snow more heat will be absorbed. It is true that global sea levels will increase, (A), but this will not directly impact the temperature. Climate will change, (E), but that is a result in the increase of temperature, not a cause. (C) and (D) are both not true.

7. E

Habitat destruction, the clear-cutting of forests especially, is the greatest threat to terrestrial species. The rest are factors but are not as prevalent.

8. D

Endangered, (A), and threatened, (C), species are species that exist in low numbers. Extinct species, (B), no longer exist. Keystone species, (E), are vital to the health of the ecosystem in some way, and non-native species never fill such a role, leaving choice (D) as the correct answer.

9. A

It would be nice if either (B) or (C) were the correct choices, but that is not the point that Garret Hardin was trying to make when he wrote his article titled "The Tragedy of the Commons."

10. C

The World Bank, (C), provides financial support to developing nations. The EPA, (D), and Endangered Species Act, (E), both deal exclusively with domestic concerns, and the Kyoto Protocol, (B), was an international agreement relating to greenhouse gases, not an agency able to provide direct aid. The U.S. Congress, (A), oversees the projects that the World Bank takes on.

ANSWER TO FREE-RESPONSE QUESTION

(A) Wildlife corridors provide areas that link one area suitable for wildlife to another. They are essential where development isolates populations of species.

(B) The National Environmental Protection Act (NEPA) requires an environmental impact statement for any area that is using federal funds. Potential damage must be assessed. The Clean Water Act, the Clean Air Act, and the Endangered Species Act could all be worked into this answer.

(C) Large mammals, such as bears, mountain lions, and wolves, are the intended beneficiaries of wildlife corridors. Of course, many other animals will benefit.

(D) A few characteristics that affect the survival of a species involve specialization in some way, such as eating only one food source or requiring very specific climate conditions. A few others are elaborate mating rituals that require space and time, low reproductive success, living on islands, and needing large areas to live and feed.

| Part Four |

PRACTICE TESTS

HOW TO TAKE THE PRACTICE TESTS

The next section of this book consists of practice tests. Taking a practice AP exam gives you an idea of what it's like to answer these test questions for a longer period of time, one that approximates the real test. You'll find out which areas you're strong in, and where additional review may be required. Any mistakes you make now are ones you won't make on the actual exam, as long as you take the time to learn where you went wrong.

The two full-length practice tests in this book each include 105 multiple-choice questions and four free-response (essay) questions. You will have 80 minutes for the multiple-choice questions, a 10-minute reading period, and 90 minutes to answer the free-response questions. Before taking a practice test, find a quiet place where you can work uninterrupted for three hours. Time yourself according to the time limit at the beginning of each section. It's okay to take a short break between sections, but for the most accurate results you should approximate real test conditions as much as possible. Use the 10-minute reading period to plan your answers for the free-response questions, but don't begin writing your responses until the 10 minutes are up.

As you take the practice tests, remember to pace yourself. Train yourself to be aware of the time you are spending on each question. Try to be aware of the general types of questions you encounter, as well as being alert to certain strategies or approaches that help you to handle the various question types more effectively.

After taking a practice exam, be sure to read the detailed answer explanations that follow. These will help you identify areas that could use additional review. Even when you've answered a question correctly, you can learn additional information by looking at the answer explanation.

Finally, it's important to approach the test with the right attitude. You're going to get a great score because you've reviewed the material and learned the strategies in this book.

Good luck!

PRACTICE TEST 1

1. At which type of plate boundary is new crust formed?

   (A) Convergent
   (B) Transform
   (C) Divergent
   (D) Subduction
   (E) Hot spot

2. Which of the following terms describes the ability of lava to flow?

   (A) Viscosity
   (B) Permeability
   (C) Porosity
   (D) Texture
   (E) Composition

3. A rock that has been changed due to heat and pressure is an example of what kind of rock?

   (A) Igneous
   (B) Metamorphic
   (C) Sedimentary
   (D) Intrusive
   (E) Extrusive

4. Which of the following terms describes the addition of water to the atmosphere by plants?

   (A) Evaporation
   (B) Precipitation
   (C) Condensation
   (D) Infiltration
   (E) Transpiration

5. Which of the following is the force responsible for driving the water cycle?

   (A) Gravity
   (B) The Sun
   (C) Earth's rotation
   (D) Earth's revolution
   (E) Temperature differences

6. Why is the carbon cycle important to humans?

   (A) It helps restore the fertility of the soil.
   (B) Carbon is responsible for metabolic processes within cells.
   (C) It is a major factor in distributing heat and energy around Earth.
   (D) It is a major factor in climate regulation.
   (E) The cycle stimulates the growth of algae.

7. Which of the following describes the impact of human sources on the carbon cycle?

   (A) Net increase of carbon dioxide in the atmosphere
   (B) Net decrease of carbon dioxide in the atmosphere
   (C) Net increase of carbon in the atmosphere
   (D) Net decrease of carbon in the atmosphere
   (E) No impact

GO ON TO THE NEXT PAGE ⟩

8. Which of the following best describes the role played by nitrogen-fixing bacteria in the nitrogen cycle?

   (A) They create organic nitrogen.
   (B) They convert nitrates to ammonia.
   (C) They convert ammonia to nitrates.
   (D) They convert nitrogen to molecular nitrogen.
   (E) They alter the organic nitrogen.

9. Which of the following denotes atmospheric nitrogen?

   (A) NO_3^-
   (B) NH_4^+
   (C) N_2O
   (D) N_2
   (E) NO_2^-

10. The ozone layer is located in which layer of the atmosphere?

    (A) Troposphere
    (B) Stratosphere
    (C) Mesosphere
    (D) Thermosphere
    (E) Exosphere

11. Which gases make up the majority of Earth's atmosphere?

    (A) Carbon dioxide and oxygen
    (B) Hydrogen and helium
    (C) Hydrogen and oxygen
    (D) Water vapor and carbon dioxide
    (E) Nitrogen and oxygen

Use the graph below to answer the following two questions.

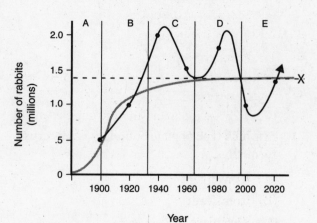

12. The horizontal line labeled x represents which of the following?

    (A) A population crash
    (B) The carrying capacity of the population
    (C) The exponential growth of the population
    (D) The population overshooting the carrying capacity
    (E) The lag phase of the population

13. Which of the following parts of this graph shows the exponential growth of the population?

    (A) A
    (B) B
    (C) C
    (D) D
    (E) E

14. Which ecosystem produces the most biomass?

    (A) Desert
    (B) Savanna
    (C) Tundra
    (D) Tropical rain forest
    (E) Temperate deciduous forest

GO ON TO THE NEXT PAGE ▷

15. The least amount of energy is available at which tropic level?

    (A) Tertiary consumers
    (B) Secondary consumers
    (C) Primary consumers
    (D) Primary producers
    (E) Secondary producers

16. Which of the following are consumers that feed at all trophic levels?

    (A) Carnivores
    (B) Parasites
    (C) Herbivores
    (D) Secondary consumers
    (E) Fungi

17. What is the name of the process in which members of a population that are best suited for environmental conditions survive and reproduce?

    (A) Evolution
    (B) Adaptation
    (C) Natural selection
    (D) Genetic diversity
    (E) Random mutations

18. The island of Madagascar has a huge number of species not found on the nearby African mainland. The difference between the species found on Madagascar and those on the mainland is most likely the result of

    (A) temporal isolation.
    (B) geographic isolation.
    (C) behavioral isolation.
    (D) sympatric speciation.
    (E) mechanical isolation.

19. A jack rabbit is well suited for a desert environment because of which of the following adaptations?

    (A) Strong hind legs
    (B) Long whiskers
    (C) Sense of smell
    (D) Long ears
    (E) Speed

20. Which of the following is NOT an environmental factor that can influence fertility or survivability?

    (A) Physiological stresses
    (B) Predation
    (C) Competition
    (D) Genetics
    (E) Chance

21. Which of the following accounts for most of the NO_x emissions in the United States?

    (A) Electric utilities
    (B) Industrial sources
    (C) Processing of petroleum
    (D) Transportation
    (E) Chemical processing

22. Which of the following is a major component in acid rain?

    (A) Sulfur dioxide
    (B) Sulfur trioxide
    (C) Sulfuric acid
    (D) Nitrogen dioxide
    (E) Carbon dioxide

GO ON TO THE NEXT PAGE

23. Which of the following is a nonpoint source of water pollution?

   (A) Drainpipes
   (B) Ditches
   (C) Sewer outfall
   (D) Atmospheric deposition
   (E) Leaking gasoline tank

24. Which of the following is NOT true of particulate matter pollution?

   (A) Particulate matter includes dust, ash, soot, lint, smoke, pollen, and spores.
   (B) Particulate matter is solid or liquid particles suspended in a gas.
   (C) Particulate matter can directly enter the lungs, damaging lung tissue.
   (D) Particulate matter is one of the least obvious forms of air pollution.
   (E) Particulate matter can cause or promote cancer.

25. Which of the following is a hazard of thermal pollution?

   (A) Increased evaporation
   (B) Pollution of public drinking water
   (C) Disruption of aquatic life
   (D) Eutrophication
   (E) Increase in algal content

26. Which of the following is a source of thermal pollution?

   (A) Magmatic activity
   (B) Geothermal springs
   (C) Agriculture
   (D) Increased organic activity
   (E) Power plant cooling systems

27. The pH scale is a measurement of the concentration of

   (A) H_3O^+ ions.
   (B) H^+ ions.
   (C) NaOH.
   (D) HCl.
   (E) OH^- ions.

28. What happens to the oxygen solubility in water as the temperature increases?

   (A) It increases.
   (B) It decreases.
   (C) It stays the same.
   (D) It depends on the salinity of the water.
   (E) It depends on the depth of the water.

29. What is typically TRUE of water temperatures?

   (A) They are more stable than air temperatures.
   (B) They are less stable than air temperatures.
   (C) They always decrease with depth.
   (D) They always increase with depth.
   (E) They change with air temperature.

30. Water that has been polluted by excess nutrients may have what type of problems?

   (A) Decrease in photosynthesis for aquatic plants
   (B) Increase in photosynthesis for aquatic plants
   (C) Too much oxygen for fish
   (D) Reduction in water's ability to hold dissolved oxygen
   (E) Increase in aerobic bacteria

GO ON TO THE NEXT PAGE

31. What impact could an influx of agricultural waste or manure have on an aquatic ecosystem?

(A) Increase in dissolved oxygen

(B) Excessive growth of algae

(C) Reduction in photosynthesis

(D) Lower BOD (biological oxygen demand)

(E) Thermal shock

32. Buildings or statues made of which materials are particularly susceptible to acid rain?

(A) Granite

(B) Basalt

(C) Wood

(D) Marble

(E) Concrete

33. The pH of normal, or average, rainfall is about

(A) 2.3.

(B) 5.7.

(C) 7.1.

(D) 8.5.

(E) 8.8.

34. Which type of plants requires more pesticides?

(A) Flowering plants

(B) Vegetables

(C) Native plants

(D) Non-native high-yield crops

(E) Agricultural plants

35. What impact does the introduction of thermal pollution have on an aquatic ecosystem?

(A) Reduces the ability of the water to hold dissolved carbon dioxide

(B) Increases the ability of the water to hold dissolved carbon dioxide

(C) Reduces the ability of the water to hold dissolved oxygen

(D) Increases the ability of the water to hold dissolved oxygen

(E) No impact on dissolved carbon dioxide or dissolved oxygen

36. Methane recovery is a positive trend in which of the following methods of solid waste disposal?

(A) Ocean dumping

(B) Open dumps

(C) Sanitary landfills

(D) Exportation of waste

(E) Incineration

37. Which of the following accounts for the greatest percentage of waste generated in homes in the United States?

(A) Metals

(B) Plastics

(C) Food

(D) Glass

(E) Paper and paperboard

GO ON TO THE NEXT PAGE

38. What is municipal waste?

    (A) Waste products from industry
    (B) Waste products from mining operations
    (C) Waste from household and commercial sites
    (D) Crop residue and animal manure
    (E) Construction debris

39. How is most municipal waste in the United States disposed?

    (A) Open dumps
    (B) Land fills
    (C) Incineration
    (D) Dumped in the ocean
    (E) Exported to other regions

40. Which of the following indoor air pollutants has the most significant impact on human health?

    (A) Radon
    (B) Formaldehyde
    (C) Cigarette smoke
    (D) Nitrogen oxides
    (E) Chloroform

41. Long-term exposure to radon gas can lead to

    (A) dizziness.
    (B) headaches.
    (C) sick-building syndrome.
    (D) lung cancer.
    (E) fatigue.

42. Which of the following is an example of a chronic condition that could be caused by exposure to air pollution?

    (A) Anemia
    (B) Emphysema
    (C) Lead poisoning
    (D) Decrease in oxygen levels
    (E) Skin cancer

43. It has been estimated that nearly half of the cancers caused by outdoor air pollution are the result of

    (A) industrial sources.
    (B) acid rain.
    (C) photochemical smog.
    (D) mobile sources, such as cars and trucks.
    (E) smoking.

44. When considering the risk assessment of different biologic agents, the pathogenicity of the agent must be considered. What does this refer to?

    (A) Geographic location of the source
    (B) The amount of the agent that is needed to produce an infection
    (C) Disease incidence and severity
    (D) Ability of the agent to survive over time
    (E) Number of infectious organisms per unit volume

45. A weakening of the trade winds and a warming of the surface ocean waters may indicate which of the following?

    (A) La Niña
    (B) El Niño
    (C) Rising sea levels
    (D) Global climate change
    (E) An increase in stratospheric ozone

46. How long is the average residence time of a molecule of carbon dioxide in the atmosphere?

    (A) 1 year
    (B) 10 years
    (C) 100 years
    (D) 1,000 years
    (E) 10,000 years

GO ON TO THE NEXT PAGE ⟩

47. A rapid increase in CO_2 in the atmosphere coincided with which of the following events?

(A) The Industrial Revolution

(B) The Kyoto Protocol

(C) Thinning of the ozone layer

(D) Invention of the automobile

(E) Establishment of the Environmental Protection Agency

48. Which of the following statements about the absorption of carbon dioxide is MOST accurate?

(A) Wetlands and forests absorb less carbon dioxide than agricultural land.

(B) Wetlands and forests absorb more carbon dioxide than agricultural land.

(C) Converting natural lands to agricultural lands has no impact on the amount of carbon dioxide absorbed.

(D) Land developed for homes and businesses absorb more carbon dioxide than agricultural land.

(E) Carbon dioxide is not absorbed by agricultural lands.

49. Sedimentary rocks are a major sink for

(A) oxygen.

(B) methane.

(C) sulfur.

(D) carbon.

(E) hydrogen.

50. What is the residence time for methane in the atmosphere?

(A) 1 year

(B) 12 years

(C) 130 years

(D) 250 years

(E) 1,000 years

51. Which organisms created the methane found in the early atmosphere?

(A) Cyanobacteria

(B) Blue-green algae

(C) Methanogenic bacteria

(D) Plants

(E) Early fish

52. Which of the following is currently the MOST significant source of methane?

(A) Cultivation of rice

(B) Wastewater treatment

(C) Coal mining

(D) Natural gas processing

(E) Landfills

53. In which layer of the atmosphere are most air pollutants located?

(A) Stratosphere

(B) Troposphere

(C) Exosphere

(D) Mesosphere

(E) Thermosphere

54. A La Niña event will bring warm winters to which part of the United States?

(A) Northeast

(B) South

(C) Northwest

(D) Southeast

(E) Midwest

55. The most dangerous radiation that stratospheric ozone protects life on Earth from is

(A) UVA radiation.

(B) UVB radiation.

(C) UVC radiation.

(D) ions from the ionosphere.

(E) microwaves.

GO ON TO THE NEXT PAGE

56. CFCs in the atmosphere decompose from exposure to which of the following?

    (A) Stratospheric ozone
    (B) Atomic chlorine
    (C) Ultraviolet radiation
    (D) Sunlight
    (E) Carbon dioxide

57. Which of the following, found in fire extinguishers and naturally released by marine phytoplankton, has more of an impact on the depletion of stratospheric ozone than chlorine?

    (A) Methane
    (B) Chromium
    (C) Hydrogen
    (D) Bromine
    (E) Benzene

58. How much warmer is Earth today than it was during the last Ice Age?

    (A) 1°–3°C
    (B) 5°–9°C
    (C) 12°–15°C
    (D) 20°–25°C
    (E) 35°C

59. Why might global temperatures be more extreme during the night than during the day?

    (A) Evaporation proceeds faster at night.
    (B) Fossil fuel consumption increases at night.
    (C) More solar radiation is absorbed during the day.
    (D) Heat is trapped near the surface at night.
    (E) Earth is closer to the Sun at night.

60. What would be the average temperature on Earth if the Greenhouse Effect did not exist?

    (A) 5°F
    (B) 35°F
    (C) 65°F
    (D) 100°F
    (E) 135°F

61. Which of the following would be a result of higher temperatures on Earth?

    (A) A decrease in precipitation worldwide
    (B) An increase in precipitation worldwide
    (C) A decrease in precipitation at lower latitudes and an increase in precipitation at higher latitudes
    (D) An increase in precipitation at lower latitudes and a decrease in precipitation at higher latitudes
    (E) An increase in evaporation worldwide, but the precipitation rates would not change

62. How does global warming impact the spread of infectious disease?

    (A) Global warming decreases the amount of infectious diseases.
    (B) Global warming has no impact on the spread of infectious disease.
    (C) Rising temperatures allow disease-carrying mosquitoes to move into new areas.
    (D) Rising temperatures kill the larvae of disease-carrying mosquitoes.
    (E) Global warming increases moisture, which drowns disease-spreading mosquitoes.

GO ON TO THE NEXT PAGE ⇒

63. Which of the following is the MOST effective way to reduce global warming?

(A) Curbing emissions of greenhouse gases

(B) Reducing CFCs

(C) Using clearer fuels

(D) Consuming energy more efficiently

(E) Recycling

64. Which of the following energy conversions is represented by photosynthesis?

(A) Solar to electrical

(B) Chemical to mechanical

(C) Solar to chemical

(D) Chemical to electrical

(E) Chemical to heat

65. Which form of energy is represented by X-rays?

(A) Chemical

(B) Radiant

(C) Electrical

(D) Potential

(E) Mechanical

66. Which type of energy is usually a low-quality by-product of energy transformations?

(A) Chemical

(B) Radiant

(C) Nuclear

(D) Heat

(E) Mechanical

67. Which energy source does NOT come from Earth?

(A) Chemical

(B) Radiant

(C) Nuclear

(D) Heat

(E) Mechanical

68. The conversion of low-quality energy to high-quality energy violates which physical law?

(A) First law of thermodynamics

(B) Second law of thermodynamics

(C) Zeroth law of thermodynamics

(D) Coulomb's law

(E) Law of inertia

69. Worldwide, most of the available freshwater is used for which of the following?

(A) Oil and gas production

(B) Power plant cooling

(C) Industrial processing

(D) Public drinking water

(E) Agriculture

70. What consequence of overusing groundwater would affect areas that are distant from the point of use?

(A) Groundwater contamination

(B) Saltwater intrusion

(C) Subsidence

(D) Depletion of stream flow

(E) Aquifer replenishment

71. Bioventing and thermal desorption are techniques sometimes used in which of the following processes?

(A) Tapping groundwater

(B) Transferring watersheds

(C) Remediation

(D) Curbing water waste

(E) Minimizing water losses from irrigation

GO ON TO THE NEXT PAGE

72. Which environment's soil will be weak in humus?

    (A) Tropical rain forest
    (B) Deciduous forest
    (C) Grassland
    (D) Desert
    (E) Coniferous forest

73. Which of the following factors does NOT influence a soil's ability to hold water, nutrients, and air?

    (A) Soil texture
    (B) Soil composition
    (C) Soil porosity
    (D) Soil structure
    (E) Soil pH

74. Which of the following countries has the greatest biodiversity?

    (A) Iceland
    (B) Great Britain
    (C) Libya
    (D) Brazil
    (E) Canada

75. Which of the following characteristics could make a species prone to extinction?

    (A) High reproduction rate
    (B) Feed at low trophic levels
    (C) Small body size
    (D) Specialized feeding habits
    (E) Found in many places

76. Which of the following is NOT a characteristic associated with the Endangered Species Act of 1973?

    (A) It was the first legislation of its type to give legal standing to wildlife.
    (B) Species on the list cannot be hunted, collected, killed, or injured in the United States.
    (C) All commercial shipments of wildlife and wildlife products must enter or leave through nine designated ports.
    (D) Federal agencies may not fund, authorize, or carry out projects that jeopardize, threaten, modify, or destroy habitats of a species on the endangered species list.
    (E) Decisions to list or to unlist a species must be made on the basis of economics and biology.

77. Which of the following is NOT a reason necessary to preserve wild species?

    (A) Some species are keystone species in their ecosystems.
    (B) Each species must have an inherent value to us to be preserved.
    (C) Some species have economic and medicinal benefits to us.
    (D) Each species is unique and irreplaceable.
    (E) Each species is of scientific interest and has a role in its ecosystem.

78. Which of the following is an effective way of preserving the interrelationships of species in a self-sustaining manner?

    (A) Zoo
    (B) A single large, biosphere reserve
    (C) Numerous small, separate wildlife areas
    (D) Private collection
    (E) Small nature preserves

GO ON TO THE NEXT PAGE

79. Which of the following is a disadvantage of developing oil shale or tar sands for crude oil production?

    (A) The supply is not enough to meet demands.

    (B) The technology does not exist to recover oil from these types of deposits.

    (C) These sources will burn cleaner than conventional crude oil.

    (D) They yield more useful products and more energy than conventional crude oil.

    (E) They require more processing and expense to extract crude oil than conventional oil sources.

80. Most of the coal mined in the world is used for which of the following processes?

    (A) Steel production

    (B) Synthetic gasoline production

    (C) Synthetic natural gas production

    (D) Production of electricity

    (E) Chemical industry

81. Which of the following fossil fuels burns cleaner and hotter than the others?

    (A) Coal

    (B) Natural gas

    (C) Oil

    (D) Oil shale

    (E) Tar sand

82. Which of the following city arrangements controls urban growth, while providing open areas for recreation and nondestructive uses?

    (A) Concentric circle city

    (B) Sector city

    (C) Multiple nucleated city

    (D) Gentrified city

    (E) Greenbelt city

83. Which of the following urban problems is directly related to cheap gasoline prices?

    (A) Runoff and flooding

    (B) Decreased vegetation

    (C) Urban sprawl

    (D) Urban microclimates

    (E) Solid waste pollution

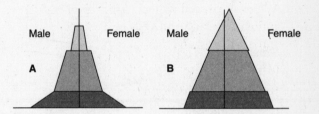

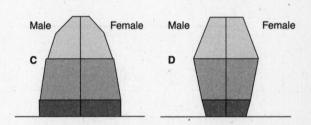

84. In the age-structure diagram above, Graph A represents which of the following countries?

    (A) Germany

    (B) India

    (C) Great Britain

    (D) United States

    (E) France

85. What is the current estimate of the size of the human population?

    (A) 6 million

    (B) 6 billion

    (C) 7 billion

    (D) 7 billion

    (E) 9 billion

GO ON TO THE NEXT PAGE

86. Which of the following countries has the lowest population?

   (A) China
   (B) Indonesia
   (C) India
   (D) United States
   (E) Canada

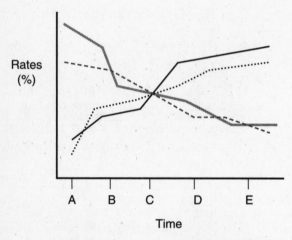

Country	Annual Rate of Population Change (%)
A	2.20
B	3.36
C	1.85
D	0.75
E	0.92

88. According to the data in the table above, which country has the shortest doubling time?

   (A) Country A
   (B) Country B
   (C) Country C
   (D) Country D
   (E) Country E

87. According to the figure above, at which time would the change in the population growth equal zero?

   (A) Time A
   (B) Time B
   (C) Time C
   (D) Time D
   (E) Time E

Country	Infant Mortality (%)
A	2.20
B	0.5
C	1.00
D	0.75
E	0.92

89. According to the data in the table above, which country has the highest fertility rate?

   (A) Country A
   (B) Country B
   (C) Country C
   (D) Country D
   (E) Country E

GO ON TO THE NEXT PAGE

90. Which of the following choices best describes the distribution of the human population worldwide?

    (A) Randomly distributed across the globe and within countries

    (B) Randomly distributed across the globe, but uniformly distributed within countries

    (C) Uniformly distributed across the globe, but clustered within countries

    (D) Clustered across the globe and clustered within countries

    (E) Clustered across the globe and uniform within countries

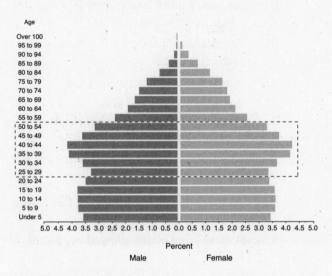

Age-Structure Diagram of the U.S. Resident Population as of July 1, 2000

Courtesy of U.S. Census Bureau, available at www.census.gov/population/www/projections/np_p2.gif

91. Look at the age-structure diagram shown above. What happened in the region highlighted by the outlined box?

    (A) Females outnumber males 2:1.

    (B) The dependency load is small.

    (C) The males outnumber females 2:1.

    (D) There was a "baby bust" or decline in births with this cohort.

    (E) There was a "baby boom" or growth spurt with this cohort.

92. A United Nations Foreign Aid group introduces agricultural technology into an LDC or third-world country to increase food productivity. What is a likely consequence?

    (A) Decrease in the environmental degradation in that country

    (B) Decrease in the carrying capacity of that country

    (C) Increase in the carrying capacity of that country

    (D) No effect on the carrying capacity of that country

    (E) Reduction in that country's birth rate

93. Which of the following is NOT a difficulty in defining the carrying capacity for the human population?

    (A) Humans live within geopolitical not environmental borders.

    (B) Humans can change the carrying capacity through technology.

    (C) When humans exceed the carrying capacity of an environment, they merely die off.

    (D) There is no logistical model that scientists can agree on to estimate the carrying capacity of the human population.

    (E) When humans exceed the carrying capacity of an environment, others can supply aid and maintain the population despite environmental degradation.

94. For humans, carrying capacity would more likely be equated with which of the following?

    (A) Survival

    (B) Standard of living

    (C) Death rates

    (D) Intrinsic rate of reproduction

    (E) The r-strategies

GO ON TO THE NEXT PAGE

95. Which of the following activities will likely have the highest external cost?

    (A) Grazing on private lands
    (B) Erecting fences on private lands
    (C) Farming on private lands
    (D) Mining on public lands
    (E) Logging on private lands

96. Which of the following are NOT approaches to saving land from the consequences of mining, logging, grazing, urban projects, and farming?

    (A) Preservation
    (B) Remediation and restoration
    (C) Ecosystems services
    (D) Sustainable land-use strategies
    (E) Aquaculture

97. What is a major limitation to using cost-benefit analysis in environmental economics?

    (A) Exact costs of the activity are often unknown.
    (B) The market value of environmental benefits is known.
    (C) The market value of environmental benefits is often unknown.
    (D) Benefits are collected instantly and up front.
    (E) It is difficult to analyze the costs and benefits if the environmental project was not done.

98. Which of the following is the first U.S. law to regulate commerce in fish and wildlife?

    (A) Endangered Species Act (ESA)
    (B) National Environmental Policy Act
    (C) Convention on International Trade in Endangered Species (CITES)
    (D) Lacey Act
    (E) Magnuson-Stevens Act

99. Which of the following is NOT a contributor to acid rain?

    (A) Coal fired power plants
    (B) Oil and natural gas refineries
    (C) Nitrogen oxide
    (D) Sulfur dioxide
    (E) CFCs

100. Which of the following is NOT associated with UNESCO?

    (A) Protection of cultural property in the event of war
    (B) Prohibiting and preventing the illegal import, export, and ownership transfer of cultural property
    (C) Safeguarding property of outstanding universal value
    (D) World cultural heritage list
    (E) Preserving historical sites

101. Which of the following is a provision of the Surface Mining Control and Reclamation Act of 1977?

    (A) All deep shaft mines must be filled in when they are shut down.
    (B) Strip mining is not allowed on public lands.
    (C) Mining companies must replant vegetation on land that was strip-mined.
    (D) Mining companies must replant vegetation on land that was deep-mined.
    (E) Mining companies must clean up the water supply on land that was strip-mined.

GO ON TO THE NEXT PAGE ⇨

102. The Comprehensive Environmental Choice, Compensation, and Liability Act is also known as _____ and is administered by _____.

(A) Toxic Waste Clean Up Act; U.S. Bureau of Land Management

(B) Superfund; U.S. Environmental Protection Agency

(C) Clean Air Act; U.S. Environmental Protection Agency

(D) Clean Water Act; U.S. Coast Guard

(E) Superfund; Federal Emergency Management Agency

103. Which of the following statements is TRUE about a conservation easement?

(A) The landowner sells the land to a government.

(B) The conservation easement ends when the landowner dies.

(C) Private organizations may not participate in conservation easements.

(D) The landowner sells the rights to development of part or all of the land.

(E) If the property gets sold, the conservation easement becomes null and void.

104. Which of the following types of remediation would be categorized as destruction or transformation?

(A) Brownfield capping

(B) Bioventing

(C) Stabilization or solidification

(D) Phytoremediation

(E) Thermal desorption

105. Land acquisition is an important part of which land use strategy?

(A) Preservation

(B) Reclamation

(C) Remediation

(D) Sustainability

(E) Mitigation

IF YOU FINISH BEFORE TIME IS CALLED, YOU MAY CHECK YOUR WORK ON THIS SECTION ONLY. DO NOT TURN TO ANY OTHER SECTION IN THE TEST.

STOP

FREE-RESPONSE QUESTIONS

1. Invasive species are one of the most serious threats to indigenous species. One particularly damaging example is that of the cane toad (*Bufo marinus*).

   Cane toads were introduced to Australia in the mid-1930s to control the cane beetle, a pest that severely devastated sugar cane crops. It was believed that cane toads had been used successfully for this purpose in the Caribbean. Unfortunately, the Australian cane beetles stay high on the stalks of the plants, and the toads could not jump high enough to catch them. Over the past 80 years, cane toads have multiplied rapidly and have spread throughout much of Australia. They are responsible for the reduction of several species of wildlife and the deaths of many domestic pets because the secretions from the skin of the toads are poisonous, among other reasons.

   Today, ecologists are studying the impact of the toads on the natural environment of Australia. By understanding the way in which the toads interact with other forms of life, ecologists hope they can find a way to eradicate this terrible pest.

   (A) (2 points) Specify two characteristics of an introduced species that cause it to be considered a pest.

   (B) (3 points) Describe two physical or behavioral adaptations that allow introduced species to gain competitive advantage over populations of a native species.

   (C) (1 point) Discuss a strategy to combat the cane toad problem.

   (D) (2 points) Name one other introduced species and its location in the world that has become a pest.

2. You have been contacted to do an ecological survey of an island off the coast of France. You need to survey the wildlife, vegetation, and climatic history. The island was settled hundreds of years ago, and there are no records of the original vegetation. Today, ground cover on the island is divided between farm fields and undeveloped areas. The undeveloped areas are located on the heights above the ocean, which is covered in a variety of grasses and shrubs, with a scattering of isolated trees. The mammalian life consists of small rodents such as hedgehogs and field mice. Average annual temperatures are warmer than would be expected at such a northern latitude due to the influence of the Gulf Stream.

   (A) (3 points) What survey method would you use to quantify the ground cover on the island?

   (B) (3 points) What survey method would you use to quantify the mammalian life?

   (C) (2 points) How can you determine the climatic records prior to development in the absence of written records?

   (D) (3 points) Explain how ocean currents in the vicinity affect the climate of the island.

3. The ozone layer in the stratosphere is critical for life on Earth because it filters harmful ultraviolet rays of the Sun. In the 1970s, atmospheric scientists recognized through satellite imagery the presence of a growing hole in the ozone layer over the South Pole. In the early 1990s, a smaller hole developed over the North Pole. The seriousness of the problem was immediately understood by the scientific community, and international legislation addressed the issue. Recent studies show that it is recovering very significantly, to the point where is it expected to be fully restored by the end of the century.

   (A) (2 points) Describe the function of ozone in the stratosphere.

   (B) (3 points) Explain the chemical interactions that are responsible for deteriorating the ozone layer in the stratosphere.

   (C) (2 points) Name and outline the international legislation responsible for assisting the recovery of the ozone layer.

   (D) (3 points) Ozone is a pollutant in the troposphere (lower atmosphere). Where does it originate? Name one health problem associated with tropospheric ozone.

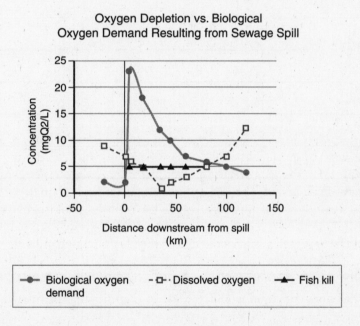

Oxygen Depletion vs. Biological
Oxygen Demand Resulting from Sewage Spill

4. The graph above shows oxygen depletion resulting downstream from a sewage spill.

   (A) (2 points) On the graph, mark the point at which the oxygen has recovered from the sewage spill. What is the concentration at this point?

   (B) If nutrients such as nitrates and phosphates are included in a sewage spill in a lake, what are the consequences to the vegetation and animal life in the river?

   (C) Turbidity, caused by sedimentation in the river from land disturbance in the surrounding areas, also has devastating consequences on fish life. Describe two of these problems.

   (D) The "dead zone" in the Gulf of Mexico is a large zone of hypoxia caused by excessive decomposition of algae. Explain why there is so much algae.

ANSWERS AND EXPLANATIONS

1. C

Convergent plate boundaries are areas where plates come together. This eliminates (A) and (D) as possible answers because a subduction zone is a particular type of convergent boundary. Students may think that (E) is correct because new crust is formed as a result of hot spot volcanoes. However, these are not plate boundaries.

2. A

Porosity and permeability are ways to characterize the pore space within sediment or sedimentary rocks, so answers (B) and (C) are incorrect. Texture and composition are two ways to classify igneous rocks, eliminating (D) and (E) as choices.

3. B

Igneous rocks form from magma or lava hardening, which eliminates (A) as a choice. An igneous rock that forms inside Earth's surface is intrusive while an igneous rock that forms on the surface of Earth is extrusive, eliminating choices (D) and (E). Sedimentary rocks form as sediment is compacted and cemented together, making (C) an incorrect choice.

4. E

Choice (A) is a tempting answer because we know that water is added to the atmosphere by evaporation. However, that applies to bodies of water, not plants. Choices (B) and (C) are wrong because condensation and precipitation occur to water already in the atmosphere. And choice (D), infiltration, is what happens to precipitation when it falls to Earth's surface.

5. B

Earth's rotation on its axis and revolution around the Sun are responsible for such phenomena as the seasons, the day, the year, and eclipses. This eliminates (C) and (D). Gravity would not cause the water to evaporate, so (A) is eliminated. Though temperature differences seem like a valid answer, even temperature differences are caused by the Sun, making (B) the correct answer.

6. D

Choice (A) describes the nitrogen cycle. Water is responsible for metabolic processes within soil, eliminating choice (B). Choice (E) describes the phosphorous cycle. Choice (C) describes the water cycle, leaving choice (D). Carbon, in the form of CO_2, has a major influence on temperature.

7. A

The burning of fossil fuels releases carbon dioxide into the air, which will increase the amount of carbon dioxide, not elemental carbon (which would be graphite or diamond), into the atmosphere.

8. C

The nitrogen cycle is a complicated process that involves converting certain chemicals into nitrates so they can be used by organisms and plants in the soil. The nitrogen-fixing bacteria specifically change ammonia into nitrates.

9. D

Nitrogen does exist in the atmosphere as N_2O—choice (C)—but this is not atmospheric nitrogen, it is nitrous oxide.

10. B

While ozone may be found in other layers of the atmosphere, the ozone layer is found in the second layer above Earth's surface, the stratosphere.

11. E

Nitrogen and oxygen make up 99 percent of the volume of clean, dry air. The other gases listed here together with others make up the remaining 1 percent.

12. B

The carrying capacity, (B), is when rapid population growth has slowed down and unutilized resources diminished, thereby causing the population to reach a limit of growth. Populations do not stop growing completely when they reach the carrying capacity, but can sometimes overshoot, (D), or diminish around the carrying capacity. A population crash, (A), is shown as a steep drop in numbers. The opposite, exponential growth, (C), is a dramatic climb in the numbers. When a population is first starting out, it may take a while for the numbers to grow; this is known as the lag phase, (E).

13. B

Exponential growth should be an area on the graph where the population increases quite rapidly in a relatively short period of time.

14. D

For an ecosystem to produce a large amount of biomass, it must contain several plants and animals and have a warm, wet environment. This best describes the tropical rain forest.

15. A

There are no secondary producers, eliminating answer (E). Primary producers, (D), have the most energy available, and the amount of energy decreases as one moves up the energy pyramid, making (A) the correct answer.

16. B

Parasites, (B), can eat at any level. Carnivores, (A), only eat meat. Herbivores, (C), eat producers only.

Secondary consumers, (D), by definition eat other consumers, so cannot feed on producers. Fungi, (E), are decomposers.

17. C

Evolution, (A), is the gradual change over time of a species. Adaptation, (B), is when a group of organisms change a bit to fit in with their environment. Genetic diversity, (D), and random mutations, (E), take place on a smaller level.

18. B

Make sure to read the question carefully. It is not asking why the island of Madagascar has such a high number of species; it is asking why it has such a high number of native species that are not found on the continent. Therefore, you can immediately eliminate choices (A), (C), and (E), which are types of reproductive isolation. Reproductive isolation is only an issue when species are in the same area. Choice (D) likewise refers to speciation events that occur in the same geographic area. All of these probably help explain why there are so many species in Madagascar, but only choice (B) addresses why the island might be different than the mainland.

19. D

To survive in the hot desert environment, animals need some way to regulate body temperature. The long ears of a jack rabbit provide it with that natural cooling mechanism.

20. D

Genetics may play a role but under the specific daily conditions in the ecosystem, it is not as big a role as the other factors and is not strictly environmental. An animal's fertility and survivability within an ecosystem does depend on factors like stresses, predation, competitions, and plain old chance.

21. D

Industrial sources, (B), contribute 12 percent, electric utilities, (A), contribute 25 percent, and together petroleum, (C), and chemical, (E), processing contribute 9 percent. Transportation sources, such as nonroad vehicles and on-road vehicles, contribute 54 percent of the NO_x emissions.

22. C

Sulfur dioxide, (A), is a colorless, corrosive gas that is directly damaging to plants and animals. This gas can be further oxidized to sulfur trioxide, (B), that reacts with water to form sulfuric acid. Nitrogen dioxide, (D), is a component in photochemical smog, and carbon dioxide, (E), is a greenhouse gas.

23. D

Drainpipes, (A), ditches, (B), sewer outfalls, (C), and a leaking gasoline tank, (E), are all identifiable point sources of water pollution. Air currents carrying pollution can carry it for great distances before the pollution is reintroduced to the environment on the ground by precipitation.

24. D

Particulate matter is possibly the most obvious form of air pollution. Its deleterious effects to humans and the environment can take many forms. For example, it often reduces visibility. It can also leave windows and cars dirty. By definition, particulate matter is solid or liquid particles suspended in a gas, (B). Dust, ash, soot, lint, smoke, pollen, and spores are all included, (A). Particles that are less than 2.5 mm in size are among the most dangerous because they can directly enter the lungs, damaging lung tissue, (C). They are also carcinogenic, meaning that they cause or promote cancer, (E).

25. C

Thermal pollution, or the addition of hot water to an aquatic environment, does not increase evaporation, (A), pollute a public drinking water supply, (B), or increase the amount of algae, (E). Eutrophication, (D), is the natural nutrient enrichment of a lake from the runoff of plant nutrients, which also leads to an increase of algae and cyanobacteria, but thermal pollution is not involved. That only leaves (C), as high heat can have a deleterious impact on the living organisms in the water.

26. E

Although magmatic activity, (A), can heat up the water, it is not the addition of hot water to an ecosystem. Geothermal springs, (B), are caused by magmatic activity, so the same logic applies here. Agriculture, (C), does contribute to water pollution by adding organic material and pesticides to the water, but it is unlikely that it is considered thermal pollution. And an increase in organic activity, (D), may alter the ecosystem, but this is not a case of thermal pollution. Cooling systems from a power plant do often introduce heated water into an ecosystem.

27. A

Choice (A) is the only possible answer because pH is a measure of H_3O^+ ions.

28. B

Solubility will change with temperature, eliminating choice (C). The solubility of oxygen is not directly affected by salinity and depth, eliminating choices (D) and (E). This leaves increasing and decreasing as choices. As temperature increases, the oxygen solubility will decrease not increase, making choice (B) correct.

29. A

Depending on underwater conditions, water temperature can vary with depth, so (C) and (D) are not good choices. Water temperature at a certain depth is no longer affected by air temperature, eliminating choice (E). A body of water takes longer than air temperature to heat up and cool down, making water temperatures more stable, (A). This eliminates the opposite description, (B).

30. A

An addition of nutrients to water will lead to an increase in photosynthetic algae that in turn will block the sunlight to aquatic plants. This leads to a decrease in photosynthesis, (A), for these plants. As these plants die, the demand for oxygen goes up and there will be less oxygen for the fish, eliminating choice (C). Thermal pollution reduces the water's ability to hold dissolved oxygen, so choice (D) is incorrect. Organic matter increases the amount of food for aerobic bacteria, making choice (E) incorrect, too.

31. B

Adding agricultural wastes is not the addition of heat, so choice (E) is incorrect. Adding agricultural wastes would have no impact on the yield of crops, choice (D). It would also decrease dissolved oxygen amounts, not increase them, as indicated in choice (A). Parts of the aquatic ecosystem may ultimately experience a decrease in photosynthesis, (C), but only after an excessive growth of algae, (B).

32. D

Granite, (A), and basalt, (B), are formed from cooling magma and lava. Wood, (C), and concrete, (E), are not as susceptible to acid rain as is marble, (D). Marble is a metamorphic rock made from limestone. Limestone is composed of calcite that will dissolve in a weak acid.

33. B

A neutral pH is 7. Rainfall is slightly more acidic than that (with a pH of 5.7). Choice (C) is incorrect because it is basically neutral. Choice (A) is incorrect because it shows a too high a level of acidity. And because choices (D) and (E) indicate less acidity, they are also incorrect.

34. D

There is no indication that flowering plants, (A), vegetables, (B), or agricultural plants in general, (E), require any more pesticides than other plants. The application of pesticides on these plants would be up to the grower. Native plants, (C), have evolved natural adaptations to local pests, so non-native high-yield plants, (D), require more pesticides.

35. C

Thermal pollution does have an impact on the dissolved oxygen in the water, not the carbon dioxide levels, eliminating choices (A), (B), and (E). It reduces, not increases, the ability of the water to hold dissolved oxygen, eliminating choice (D).

36. C

Methane is a gas, so recovering the gas in an open dump, (B), or from the open ocean, (A), would be next to impossible. It would also be difficult to capture a gas just by exporting waste, (D). Burning waste, (E), will release the gas, but it is in a much less controlled environment, while in a sanitary landfill, pipes can be inserted to catch the methane as it is released. This makes choice (C) the best option.

37. E

According to a report issued in October 2006 by the Environmental Protection Agency, of the 240 million tons of municipal solid waste generated in the United States in 2005, paper and paperboard make up the largest portion (34 percent). Plastics, (B), and food scraps, (C), are also big contributors, at about 11 percent each. Metals, (A), make up about 7 percent, while glass, (D), makes up only about 5 percent. Other contributors include yard trimmings, which are actually the next largest contributor at 13 percent. Wood makes up about 5 percent; rubber, leather, and textiles combined make up about 7 percent; and other miscellaneous wastes make up approximately 3 percent.

38. C

Waste from industry, including mining, is called industrial waste. This eliminates choices (A) and (B). Agricultural waste includes crop residue and animal manure, (D). Construction debris, (E), is a major component of solid waste, but is not considered municipal waste.

39. B

The dumping of wastes into the ocean is illegal in the United States, eliminating choice (D). The use of open dumps is declining, especially in more urban areas. There are major environmental concerns about incinerating a large amount of waste, making choice (C) unlikely. Because exporting the material is costly and complicated, the best choice is landfills, choice (B).

40. C

Nitrogen oxides are an outdoor concern, so choice (D) can be eliminated. Radon, (A), formaldehyde, (B), and chloroform, (E), are all examples of indoor air pollutants, but they are not as common, therefore their impact is not as great, as cigarette smoke, (C).

41. D

Radon is a carcinogen that can cause lung cancer. It is a colorless, odorless gas that is the result of the radioactive decay of uranium.

42. B

Air pollution does not cause skin cancer, (E). Anemia, (A), is a lack of healthy red blood cells, usually caused by iron deficiency or blood loss, and is not related to air pollution. A decrease in oxygen levels, (D), is not a chronic condition. Lead is a solid, so lead poisoning, (C), can be a result of soil or water pollution, not usually air pollution.

43. D

Acid rain, (B), and photochemical smog, (C), are *results* of air pollution, so those choices are incorrect. Smoking is an indoor air pollutant, eliminating choice (E). Industrial sources, (A), while significant, do not contribute to the cancer link as much as mobile sources.

44. C

Choice (A) is the origin of the agent, not the pathogenicity, so (A) cannot be correct. Choice (B) refers to the infectious dose of the agent, so (B) is also incorrect. Choice (D) is the agent stability, which is incorrect, and choice (E) is the concentration of the agent, so (E) is eliminated also.

45. B

La Niña, (A), results in strengthening trade winds, so this choice is incorrect. While choices (C), (D), and (E) may result in a warming of the surface waters, they don't directly affect trade winds, so the best answer for this question is El Niño, (B).

46. C

A molecule of carbon dioxide has a residence time of approximately 100 years.

47. A

It was the Industrial Revolution, (A), that began the dramatic increase in CO_2 levels. The increase may cause more thinning of the ozone layer, (C), and it may have precipitated the need to establish the EPA, (E), and the Kyoto Protocol, (B), but these are issues that have come after the fact. And although the automobile, (D), does add more CO_2 to the atmosphere, its invention did not signal the increase in the levels.

48. B

Natural wetlands and forests do absorb more carbon dioxide than farmland. This means that choice (B) is correct. It also eliminates choices (A) and (C) as the correct answer. Choice (E) is tempting, but incorrect, for while crops do absorb less CO_2 than forests or wetlands, they do take up some carbon. This is why the practice of destroying natural habitats for farming is so detrimental to the environment. Choice (D), about developed land, is incorrect because the development of land for homes and businesses involves destroying natural vegetation that will significantly reduce the absorption of carbon dioxide.

49. D

Many sedimentary rocks, (D), have organic components that make them a sink for carbon.

50. B

Methane stays in the atmosphere for 12 years, so choice (B) is correct.

51. C

Cyanobacteria, (A), and blue-green algae, (B), were found in the early Earth, but they did not contribute methane to the atmosphere. Plants, (D), and early fish, (E), were not on Earth at an early enough time to impact the formation of the early atmosphere, nor do these organisms produce methane. The earliest contributor was methanogenic bacteria, (C).

52. E

While the cultivation of rice, (A), wastewater treatment, (B), coal mining, (C), and natural gas processing, (E), are all sources of methane, by far the most significant is the methane released from landfills, choice (E).

53. B

The troposphere, (B), is the layer of the atmosphere closest to the surface of Earth and the densest, so it traps most air pollutants as they are formed. Some air pollutants are eventually transported to the stratosphere, (A), where they can cause ill effects like ozone depletion, but the majority of pollutants remain within the troposphere. The mesosphere, (D), thermosphere, (E), and exosphere, (C), (listed in order of increasing distance from Earth's surface)

are atmospheric layers that are so far from Earth's surface that they contain few or no air pollutants.

54. D

La Niña events develop when air pressure increases over the Pacific Ocean, leading to an increase in trade winds. Its impact on the winters in the Northeast, (A), South, (B), and Midwest, (E), varies, but they are not characterized by a significant warming effect. A La Niña creates warmer than normal winters in the Southeast, (D), and colder than normal winters in the Northwest, (C).

55. C

Choices (A) and (B) are concerns but are not the most significant ultraviolet concern. Microwaves, (E), and ions, (D), are not of great concern to human health. The most significant protection that the ozone layer can provide is to protect humans from UVC radiation, (C).

56. C

The decomposition of CFCs in the atmosphere releases atomic chlorine, making choice (B) incorrect. This process will ultimately decrease the amount of stratospheric ozone, (A). While the components of sunlight, (D), and carbon dioxide, (E), do break down the CFCs, it is the ultraviolet radiation, (C), that does it specifically.

57. D

Bromine, (D), is about 50 times more effective than chlorine when it comes to depleting stratospheric ozone. Methane, (A), chromium, (B), hydrogen, (C), and benzene, (E), are not found in fire extinguishers and are therefore not the correct answers.

58. B

The last Ice Age, about 10,000 years ago, saw temperatures 5° to 9°C cooler than they are today, (B). This eliminates choice (A) as the correct answer because it is not a big enough range. Any of the larger ranges, as seen in choices (C), (D), and (E), would be too dramatic for life as we know it on Earth to survive.

59. D

Earth does not become closer or farther away from the Sun in its daily cycle, so that eliminates choice (E). Evaporation increases during the day, not at night, eliminating choice (A). However, the higher temperatures at night could be due to increased consumption of fossil fuels during the day (eliminating choice B). They release particulate matter into the atmosphere, which reflects solar radiation. Evaporation at night leads to more clouds being formed, which will trap that solar radiation so it stays close to Earth's surface. This eliminates choice (C), meaning choice (D) is correct.

60. A

The Greenhouse Effect makes the planet habitable by keeping the temperatures at a level that supports life. With no Greenhouse Effect, the average temperature would be around 5°F, (A).

61. C

Evaporation would be greater near the poles, and because of global wind patterns, these clouds would move to higher latitudes, bringing more water. This pattern indicates that choice (C) is correct.

62. C

Mosquitoes and rodents are able to expand their range as temperatures rise, spreading infectious diseases. This statement shows that choice (C) is correct.

63. A

While reducing CFCs, (B), using cleaner fuels, (C), consuming energy more efficiently, (D), and recycling, (E), are viable choices for reducing global warming in one way or another, the *best* way would be to curb the emission of greenhouse gases, (A).

64. C

In photosynthesis, chlorophyll absorbs energy from sunlight (i.e., solar energy), eliminating choices (B), (D), and (E). The energy gets stored in the chemical bonds, eliminating choice (A), of 3-carbon and ultimately 6-carbon sugars.

65. B

X-rays are radiated from the Sun and are part of radiant energy, so choices (A), (C), (D), and (E) are incorrect.

66. D

Low-quality energy is disorganized and has little potential to do work. So energies that are capable of doing work (chemical, radiant, nuclear, and mechanical) are incorrect, so choices (A), (B), (C), and (E) can be eliminated. Most energy transformations in biology yield heat energy as their by-product; this heat energy gets lost to the environment and is incapable of being harnessed to do work; therefore, choice (D) is correct.

67. B

Chemical energy is stored in various minerals and biomass on Earth, so choice (A) is incorrect. Radioactive isotopes are found within Earth's interior, which give rise to geothermal energy, so nuclear energy, (C), and heat energy, (D), are incorrect. Mechanical energy is found on Earth in the movements of air and water, so choice (E) is incorrect. Radiant energy from the Sun (sunlight) is the only form that comes from extraterrestrial sources, so choice (B) is correct.

68. B

The first law of thermodynamics deals with the conservation of matter and energy, so choice (A) is incorrect. The zeroth law of thermodynamics states that when two objects of different temperatures come into contact, they each will reach one uniform temperature, so choice (C) is incorrect. Coulomb's law, (D), deals with electrical charges, and the law of inertia, (E), deals with the motion of objects, so these are incorrect. The second law of thermodynamics states that in any energy transformation there is always a decrease in energy quality, so choice (B) is correct.

69. E

Agricultural irrigation accounts for about 69 percent of water withdrawal, so choice (E) is correct. About 23 percent of water withdrawal is for energy production—oil and gas production, (A), and power plant cooling, (B)—and industrial processing, (C). Only about 8 percent is used for public use, (D).

70. D

Overusing groundwater means taking water from the aquifers, not replenishing them, so choice (E) can be eliminated. Contamination is a problem of introducing pollutants into water, not of overusing it, eliminating choice (A). Saltwater intrusion, (B)—the concentration of salts and minerals in the water—occurs because of groundwater depletion, but this would be a local problem, not affecting areas far from the groundwater. Subsidence, (C), is the downward motion or collapse of Earth's surface that occurs because of deep voids or spaces underneath it. This is also a local problem. Because streams and rivers get a major portion of their water from groundwater, the depletion of groundwater would lower the flow of water, (D). This would significantly affect areas downstream.

71. C

Bioventing and thermal desorption are forms of remediation, (C), the process of removing contaminants from disturbed land. Bioventing is a type of transformation, changing harmful contaminants into something less dangerous as part of the clean-up of an environmentally degraded site,

while thermal desorption is a method of pollutant removal. None of the other choices are associated with contaminant removal.

72. D

Humus is a sticky brown layer of partially decomposed organic matter. It comes from decayed leaves and other organic matter and is usually found in the upper layers of soil. Forests and grasslands are rich in organic material for humification, so choices (A), (B), (C), and (E) are incorrect. Deserts, (D), have little vegetation for humification and have humus-poor soils.

73. E

The soil pH determines what types of plants can grow, not the physical nature of how nutrients, water, and air move through the soil. Therefore, choice (E) is correct. Soil composition is the proportion and type of soil particles (sand, silt, clay) that determine its texture. The soil texture, along with the structure (how the particles are held together), determines the soil's porosity (the size of the spaces by which air, nutrients, and water move through the soil). Therefore, choices (A) through (D) are incorrect.

74. D

Biodiversity is greatest in the tropical rain forest regions and tends to decrease as one moves toward the northern and southern latitudes. Iceland, Great Britain, and Canada are northern latitude countries and have fewer species, making choices (A), (B), and (E) incorrect. Because of its desert environment, Libya has fewer species, making choice (C) incorrect. Brazil is mostly tropical rain forests, which have the greatest biodiversity, making choice (D) correct.

75. D

Species tend to become extinct when they have low rates of reproduction, feed at high trophic levels, have large body sizes (need large amounts of food), and live in only one place. Because specialized feeding habits make the species less adaptable and more prone to extinction, choice (D) is correct.

76. E

The Endangered Species Act of 1973 was in fact the first legislation to give legal standing to wildlife, (A). It provided that species on the list could not be hunted, killed, injured, or collected, (B). The commercial shipments of wildlife and wildlife products were regulated through designated U.S. ports, (C). According to the act, federal agencies could not fund, authorize, or conduct projects that would damage habitats of endangered species, (D). Decisions as to whether a species is listed or unlisted as endangered must be made solely on biology, making choice (E) correct.

77. B

A species does not need to be valuable to us to deserve preservation, making choice (B) correct. Each species plays a role in its ecosystem, (A), and some are keystone species upon which the entire ecosystem depends, (E). Each species is unique and irreplaceable, so species should be preserved, (D). Some species have economic and medicinal benefits to us, (C).

78. B

Preserving the interrelationships of species requires setting aside large natural areas to maintain biodiversity and be self-sustaining. Because zoos, private collections, and small nature preserves do not have sufficient areas for self-sustaining systems, choices (A), (D), and (E) are incorrect. One large protected area (i.e., bioreserve) is better than numerous small areas (i.e., wildlife areas) because it allows sufficient room for species to move and interact. So choice (C) is incorrect, while choice (B) is correct.

79. E

Supplies of oil shale and tar sands could potentially meet crude oil demands for 40 years, so choice (A) is incorrect. Technologies do exist to extract oil from these sources, so choice (B) is incorrect. Both shale oil and tar sands contain higher amounts of sulfur and nitrogen compounds (pollutants) that must be removed first, so choice (C) is incorrect. Both shale oil and tar sands produce crude oil that is not as useful as conventional crude oil, as it has less potential energy, so choice (D) is incorrect. The extra processing required to extract oil from oil shale and tar sands requires more energy and expense than that of conventional oil. This has been the major limiting factor for using these sources, so choice (E) is correct.

80. D

The most common use of coal in the world is that it is burned to make steam for producing electricity, (D). Processing of coal for steel production, (A), is the second highest use of solid coal in the world. Although coal can be used to produce synthetic gasoline, (B), and natural gas, (C), these processes are expensive and not widely used. Oil, not coal, is the primary product used by the chemical industry, (E).

81. B

Natural gas burns the hottest and cleanest of all the fossil fuels, so choice (B) is correct. Coal is the dirtiest fossil fuel when burned, so choice (A) is incorrect. Oils burn somewhat cleaner than coal but vary (oil shales and tar sands burn dirtier than conventional oil), so choices (C) through (E) are incorrect.

82. E

Concentric circle, (A), sector, (B), and multiple nucleated cities, (C), are descriptions of the patterns by which cities grow. Gentrification, (D), is the process by which poor inner cities are redeveloped for higher social and economic activities. A greenbelt has an inner city and a ring of outer developed areas separated by large natural areas for recreation. Greenbelts, (E), check urban growth, while maintaining natural areas.

83. C

Runoff and flooding, (A), are problems associated with covering lands with building and concrete surfaces; the overall increase in runoff can overload sewage systems, thereby causing flooding. Increased development reduces vegetation, (B), as lands are covered. Urban microclimates, (D), are altered climates caused by building heat-generating sources (factories, buildings, roadways). Solid waste pollution, (E), is directly related to increases in urban populations. Urban sprawl is a low-population density, widely dispersed type of growth. Urban sprawl is directly related to cheap gasoline prices, car cultures, plentiful land, and roadway networks, so choice (C) is correct.

84. B

The shape of Graph A indicates a country with rapid population growth. Such graphs are characteristic of developing or less developed countries. Germany, Great Britain, France, and the United States are all industrialized nations, so choices (A), (C), (D), and (E) are incorrect. Because India is an LDC with rapid population growth, choice (B) is correct.

85. C

The current estimate of the human population size is 7 billion. With the current growth rate, the human population size on Earth is expected to reach 9 billion by the year 2050.

86. E

The five most populous countries in the world are China, India, Brazil, the United States, and Indonesia. Canada (E) is only the 36th most populous country in the world, far less populous than the other choices.

87. C

The equation for the change in population growth is as follows: $\Delta N_t = (B + I) - (D + E)$, where B is the birth rate, I is the immigration rate, D is the death rate, and E is the emigration rate. At times A and B, the birth and immigration rates, are greater than the death and emigration rates, so the change in population size is positive, making choices (A) and (B) incorrect. At times D and E, the birth and immigration rates are less than the death and emigration rates, so the change in population size is negative, making choices (D) and (E) incorrect. Only at time C are all the rates equal, so the net change in population size would equal zero, making choice (C) correct.

88. B

Doubling time is the number of years that it will take a population to double if the current growth rates remain constant. Doubling time is calculated by the rule of 70: Doubling time = 70 years ÷ annual rate of population change (%). Note that as the ARPC increases, the doubling time shortens. Country B has the longest ARPC and the shortest doubling time, making choice (B) correct.

89. A

Fertility rates are the number of children born per number of women (usually 1,000). Several factors affect fertility rates, including infant mortality rate, marriage age, education, child labor, and availability of birth control. Generally, in countries where infant mortality rates are high, the fertility rates are also high. In the table, Country A has the highest infant mortality rate and would likely have the highest fertility rate, making choice (A) correct.

90. D

Because the human population is not randomly nor uniformly distributed across the globe, choices (A), (B), and (C) are incorrect. In each country, populations are not uniformly distributed, so choice (E) is incorrect. Rather, populations are clustered along the coasts or in major urban areas.

91. E

Age-structure diagrams can indicate male-to-female ratios, but in this one the male-to-female ratio is about even, so choices (A) and (C) are incorrect. The dependency load is the portion of the population that is under age 15 and over age 65, which is not indicated, (B). There was no indentation in the diagram to indicate a baby bust, (D). The box clearly shows a bulge in the population that indicates a baby boom or growth spurt, so choice (E) is correct.

92. C

Because agricultural technology is more likely, not less, to degrade the environment, choice (A) is incorrect. Agricultural technology has no direct effect on the birth rate, but, if anything, it would likely increase the birth rate, not decrease it, making choice (E) incorrect. Humans can raise the carrying capacity of the environment through advances in technology, especially agriculture, so choices (B) and (D) are incorrect.

93. C

Human carrying capacity is difficult to define because humans live within geopolitical borders rather than environmental ones, so choice (A) is incorrect. Humans can also change their environment through technology, so choice (B) is incorrect. Scientists have not agreed upon an equation or logistical model that can describe human

population dynamics, making choice (D) incorrect. When humans exceed the carrying capacity of an environment, they can migrate or receive aid from other human populations, making choice (E) incorrect. Under these circumstances, they do not necessarily die off, so choice (C) is correct.

94. B

Carrying capacity in nonhuman populations is phrased in terms of survival, so choice (A) is incorrect. Intrinsic rate of reproduction and r-strategies are more applicable to nonhuman populations, so choices (D) and (E) are incorrect. Death rates are not part of the carrying capacity, so choice (C) is incorrect. For humans, carrying capacity is better thought of as standard of living. Countries with a high standard of living utilize more resources than those in countries with lower standards of living. Therefore, because those countries with high standards of living tend to have lower populations, choice (B) is correct.

95. D

External costs are costs that are not included in the price of a product or service. Choices (A), (B), (C), and (E) are incorrect because in all of these activities the cost of ownership of the land is not an external cost. Mining on public lands eliminates the cost of ownership as an internal cost. Furthermore, most use of public lands has only marginal fees, so choice (D) is correct. Of course, the environmental damage is an additional cost that is also not accounted for.

96. C

Ecosystems services are not an approach to saving land. They provide people with products—such as timber, food, or medicinal plants—or they perform important functions—such as waste decomposition, air and water purification, flood control, soil conservation, and erosion control. Preservation (A) prevents land from being used. This can be done through purchasing lands or placing legal restrictions on their use. Some private landowners will also donate their lands to conservation organizations for protection. Private organizations may help save sensitive ecological areas in developing countries by conserving those areas in exchange for eliminating debts. Remediation and restoration, (B), involve repairing damaged land. Remediation is the removal of pollutants from the disturbed land, including sediments in waterways. Restoration is the process by which land that has been altered by development, road construction, agriculture, mining, or grazing is returned to predisturbance conditions. Sustainable land-use strategies, (D), use ecological principles to guide land-use decisions. And aquaculture, or fish farming, (E), involves raising populations of fish in ponds or fenced areas of estuaries and bays.

97. C

Cost-benefit analysis compares the cost of doing a project with the monetary benefits from the project. The costs of a project, (A), are known immediately and up front, the benefits, (D), are collected gradually. The first step in doing a cost-benefit analysis is determining what would happen with and without the project, so choice (E) is incorrect. The major difficulty in doing a cost-benefit analysis in environmental economics is that it is often difficult to assess the monetary benefits because the market value of environmental aspects is often unknown, so choice (B) is incorrect, while choice (C) is correct.

98. D

The correct answer is the Lacey Act, (D), which was passed in 1900. Both the ESA, (A), and CITES, (C), do deal with commerce in wildlife and wildlife products, including fish, plants, and insects, but were passed much more recently than the Lacey Act. The National Environmental Policy Act, (B), deals with environmental impacts and led to the establishment of the EPA, eliminating choice (B).

The Magnuson-Stevens Act concerns fisheries regulations, eliminating choice (E).

99. E

Sulfur dioxide, (D), and nitrogen oxide, (C), are the two main chemical components of acid rain, so those can immediately be eliminated. Most of the anthropogenic sulfur dioxide in the atmosphere comes from burning fossil fuels, eliminating choice (A). During the refining process, sulfur is often removed from oil and natural gas and released into the atmosphere, so choice (B) is incorrect. CFCs have been implicated in depletion of the ozone layer, but do not contribute to acid rain, so choice (E) is correct.

100. E

UNESCO is the division of the United Nations that is associated with protecting global cultural heritage. UNESCO develops treaties and conventions to protect cultural heritage worldwide. UNESCO conventions include (1) protection of cultural property in the event of war; (2) prohibiting and preventing the illegal import, export, and ownership transfer of cultural property; and (3) safeguarding property of outstanding universal value, so choices (A) through (C) are incorrect. The World Heritage List is an inventory of cultural and natural property within the borders of all UNESCO member nations; this list is a mechanism for implementing the third UNESCO convention on safeguarding property, so choice (D) is incorrect. UNESCO does not directly manage historic sites, so choice (E) is correct.

101. C

Choices (A) and (D) are incorrect because, as its name states, the Surface Mining Control and Reclamation Act of 1977 sets regulations regarding strip or surface mining, not deep mining. The act states nothing about water supplies, so choice (E) is incorrect. The act regulates how mining can be done and how the environment should be treated during and after mining operations, not where the mining activities can occur, making choice (B) incorrect. The Surface Mining Control and Reclamation Act of 1977 requires mining companies to replant vegetation on land that was strip-mined, so choice (C) is correct.

102. B

The Comprehensive Environmental Choice, Compensation, and Liability Act is also known as Superfund, so choices (A), (C), and (D) are incorrect. Because the Superfund is managed by the U.S. Environmental Protection Agency, choice (E) is incorrect.

103. D

The landowner does not sell the land and retains full ownership rights, so choice (A) is incorrect. Answers (B) and (D) are incorrect because the conservation easement is attached to the land and remains in effect regardless of whether the property gets sold or inherited. A conservation easement is an agreement between a landowner and a government or private organization, so choice (C) is incorrect. In the agreement, the landowner sells the rights to development of part or all of the land to the other party in exchange for money or tax breaks, so choice (D) is correct.

104. B

Brownfield capping and stabilization/solidification are containment methods of remediation where the pollutant remains at the site, making choices (A) and (C) incorrect. Phytoremediation and thermal desorption are methods to remove the pollutant from the site using biological (plants) and physical (steam for dissolving the pollutant) means, making choices (D) and (E) incorrect. Bioventing is a transformation method where oxygen is injected into the site to improve the activities of aerobic microorganisms to chemically change the pollutant, so choice (B) is correct.

105. A

Land acquisition involves purchasing land, and it is mostly used as a strategy for preservation, (A). Reclamation, (B), and remediation, (C), are ways of fixing the land after it has been used. Mitigation strategies, (E), are ways to minimize the impact of land use on the environment, and they may even be a part of a sustainable land-use strategy, (D), which attempts to use the land wisely and ecologically.

ANSWERS TO FREE-RESPONSE QUESTIONS

1. This question is document based, so at least some of the answers are embedded in the text. Always look for them because they are the easiest and most straightforward.

   (A) The definition of pest is an organism that interferes with human activities. Characteristics such as competing against native species and poisoning pets certainly qualify something as a pest.

   (B) Characteristics that enable an introduced species to outcompete an indigenous one include multiplying easily, being comfortable in a wide range of living conditions, and eating a varied and nonspecific diet. They were also resistant to many of the pesticides that were used in the areas and had no natural predators.

   (C) One strategy is to prevent further spreading through removing or destroying every toad that appears. Government subsidies and rewards would help as incentives. Experiments have also been conducted in biological control of the population.

   (D) Examples are far too numerous to name, but grading a question like this requires research by the reading consultants to determine the accuracy of an answer. A few examples follow: Kudzu from Asia is currently seen along the sides of highways in the southern United States. Many grazing grasses on the prairies of the United States have replaced the natural grasses. Marsh grasses have been replaced by phragmites in wetlands. Purple loosestrife is seen all along roadsides of the Northeast. Asian long-horned beetle is found on street trees in New York. Some well-known aquatic examples include zebra mussels and silver carp.

2. (A) Many methods are used to quantify vegetation depending on characteristics such as height and density. One method involves dividing the study area into square-meter plots, measuring and selecting a random selection of plots in which to identify and count every plant. A second method involves stretching a transect string along a measured distance at a certain height above the ground, then identifying and counting the plants that touch the string at a predetermined interval.

   (B) Two common methods for quantifying animals involve setting live traps and identifying footprints on a board set with food to lure the animal. Students may be creative in their suggestions, such as using some kind of photography equipment with a lure. A description of tag and release may also be used. The Hine's or Zippen methods of counting animals based on what is captured as a percentage of what exists will also be acceptable here.

   (C) Past vegetation yields clues about climate. Vegetation prior to written records is established through studying pollen in sediment at the bottom of a lake. Trees produce pollen, which is very resistant. It falls on the surface of lakes and sinks to the bottom. It is collected in a core, a tube pushed in the soft sediment. As dictated by the Law of Superposition, pollen at the bottom of the core would be older than the pollen closer to the top. As climate changes, so do the types of trees present, and so the pollen that gets preserved changes through the core. This would certainly be a good answer. In areas where no lake exists or a lake has dried up in the past, it is difficult to establish the ancient vegetation in this way. If this is the case, marine microfossils, also collected in cores from the ocean bottom, are most helpful. Tree rings may also be helpful in that they are thicker when climatic conditions are optimal and thinner when conditions are less favorable.

(D) Ocean currents are largely responsible for terrestrial climate. Water's high specific heat means that water currents transport the temperature conditions of the areas from which they travel. Far northern latitudes usually produce cold weather, but the Gulf Stream currents from the equatorial Atlantic create a warmer climate here.

3. (A) The ozone layer in the stratosphere protects living things from ultraviolet rays of the sun. In humans, thinning ozone is credited with increased incidence of skin cancer. The thinning ozone is also responsible for harming living things that are exposed to sunlight all the time. An example is krill, the base of the marine food chain.

(B) Ozone molecules consist of three oxygen atoms, O_3. Naturally, ozone is constantly being created and destroyed in a steady state process. When chlorofluorocarbons (from coolants and propellants) rise in the atmosphere, UV radiation breaks them down. The liberated chlorine atom combines with the ozone and "pulls off" and bonds with an oxygen atom. This process happens again and again as each chlorine atom is recycled.

(C) The Montreal Protocol of 1987 is the legislation that was responsible for banning the use of chlorofluorocarbons. Signatories agreed to a 50 percent reduction of CFC production by 1998, but most manufacturers complied prior to the 1998 deadline as chemical manufacturers recognized the seriousness of the disappearance of the ozone layer.

(D) In the troposphere, the source of ozone is the burning of fossil fuels and mixing with atmospheric oxygen. Ozone is a secondary pollutant formed when nitrogen oxides and volatile hydrocarbons react. Ozone leads to decline in forest growth and crop yields. In humans, it is a lung irritant that may lead to chest pain, asthma, and bronchitis.

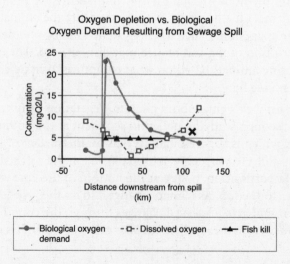

4. The question is a bit redundant, but it gets at the connection between excess nutrients, biological oxygen demand, and fish kills.

   (A) Oxygen has recovered from the sewage spill on the point marked with an X, approximately 110 km downstream from the spill. It is at this point that the oxygen level downstream of the spill equals the level upstream, which represents the baseline oxygen concentration for the unaffected stream. The oxygen concentration at this point is 9 mgO$_2$/L.

   (B) Two nutrients that are commonly found in sewage are nitrates and phosphates. These are limiting factors to certain types of algae. When they are released, the lake environment may become overgrown with algae. When algae die and decompose, they use up the oxygen, leading to the fish kill.

   (C) Turbidity caused by sedimentation fills in waterways, reduces light penetration, impairs the ability of producers to photosynthesize, changes the water temperature, and can clog gills of fish.

   (D) The algae are so plentiful because of the fertilizer runoff from the farms in the Midwest that has flowed down the Mississippi River. Algae blooms use up oxygen as they decompose, creating the hypoxic environment.

Practice Test 2 Answer Grid

1. Ⓐ Ⓑ Ⓒ Ⓓ Ⓔ
2. Ⓐ Ⓑ Ⓒ Ⓓ Ⓔ
3. Ⓐ Ⓑ Ⓒ Ⓓ Ⓔ
4. Ⓐ Ⓑ Ⓒ Ⓓ Ⓔ
5. Ⓐ Ⓑ Ⓒ Ⓓ Ⓔ
6. Ⓐ Ⓑ Ⓒ Ⓓ Ⓔ
7. Ⓐ Ⓑ Ⓒ Ⓓ Ⓔ
8. Ⓐ Ⓑ Ⓒ Ⓓ Ⓔ
9. Ⓐ Ⓑ Ⓒ Ⓓ Ⓔ
10. Ⓐ Ⓑ Ⓒ Ⓓ Ⓔ
11. Ⓐ Ⓑ Ⓒ Ⓓ Ⓔ
12. Ⓐ Ⓑ Ⓒ Ⓓ Ⓔ
13. Ⓐ Ⓑ Ⓒ Ⓓ Ⓔ

14. Ⓐ Ⓑ Ⓒ Ⓓ Ⓔ
15. Ⓐ Ⓑ Ⓒ Ⓓ Ⓔ
16. Ⓐ Ⓑ Ⓒ Ⓓ Ⓔ
17. Ⓐ Ⓑ Ⓒ Ⓓ Ⓔ
18. Ⓐ Ⓑ Ⓒ Ⓓ Ⓔ
19. Ⓐ Ⓑ Ⓒ Ⓓ Ⓔ
20. Ⓐ Ⓑ Ⓒ Ⓓ Ⓔ
21. Ⓐ Ⓑ Ⓒ Ⓓ Ⓔ
22. Ⓐ Ⓑ Ⓒ Ⓓ Ⓔ
23. Ⓐ Ⓑ Ⓒ Ⓓ Ⓔ
24. Ⓐ Ⓑ Ⓒ Ⓓ Ⓔ
25. Ⓐ Ⓑ Ⓒ Ⓓ Ⓔ
26. Ⓐ Ⓑ Ⓒ Ⓓ Ⓔ

27. Ⓐ Ⓑ Ⓒ Ⓓ Ⓔ
28. Ⓐ Ⓑ Ⓒ Ⓓ Ⓔ
29. Ⓐ Ⓑ Ⓒ Ⓓ Ⓔ
30. Ⓐ Ⓑ Ⓒ Ⓓ Ⓔ
31. Ⓐ Ⓑ Ⓒ Ⓓ Ⓔ
32. Ⓐ Ⓑ Ⓒ Ⓓ Ⓔ
33. Ⓐ Ⓑ Ⓒ Ⓓ Ⓔ
34. Ⓐ Ⓑ Ⓒ Ⓓ Ⓔ
35. Ⓐ Ⓑ Ⓒ Ⓓ Ⓔ
36. Ⓐ Ⓑ Ⓒ Ⓓ Ⓔ
37. Ⓐ Ⓑ Ⓒ Ⓓ Ⓔ
38. Ⓐ Ⓑ Ⓒ Ⓓ Ⓔ
39. Ⓐ Ⓑ Ⓒ Ⓓ Ⓔ

40. Ⓐ Ⓑ Ⓒ Ⓓ Ⓔ
41. Ⓐ Ⓑ Ⓒ Ⓓ Ⓔ
42. Ⓐ Ⓑ Ⓒ Ⓓ Ⓔ
43 Ⓐ Ⓑ Ⓒ Ⓓ Ⓔ
44. Ⓐ Ⓑ Ⓒ Ⓓ Ⓔ
45. Ⓐ Ⓑ Ⓒ Ⓓ Ⓔ
46. Ⓐ Ⓑ Ⓒ Ⓓ Ⓔ
47. Ⓐ Ⓑ Ⓒ Ⓓ Ⓔ
48. Ⓐ Ⓑ Ⓒ Ⓓ Ⓔ
49. Ⓐ Ⓑ Ⓒ Ⓓ Ⓔ
50. Ⓐ Ⓑ Ⓒ Ⓓ Ⓔ
51. Ⓐ Ⓑ Ⓒ Ⓓ Ⓔ
52. Ⓐ Ⓑ Ⓒ Ⓓ Ⓔ

53. Ⓐ Ⓑ Ⓒ Ⓓ Ⓔ
54. Ⓐ Ⓑ Ⓒ Ⓓ Ⓔ
55. Ⓐ Ⓑ Ⓒ Ⓓ Ⓔ
56. Ⓐ Ⓑ Ⓒ Ⓓ Ⓔ
57. Ⓐ Ⓑ Ⓒ Ⓓ Ⓔ
58. Ⓐ Ⓑ Ⓒ Ⓓ Ⓔ
59. Ⓐ Ⓑ Ⓒ Ⓓ Ⓔ
60. Ⓐ Ⓑ Ⓒ Ⓓ Ⓔ
61. Ⓐ Ⓑ Ⓒ Ⓓ Ⓔ
62. Ⓐ Ⓑ Ⓒ Ⓓ Ⓔ
63. Ⓐ Ⓑ Ⓒ Ⓓ Ⓔ
64. Ⓐ Ⓑ Ⓒ Ⓓ Ⓔ
65. Ⓐ Ⓑ Ⓒ Ⓓ Ⓔ
66. Ⓐ Ⓑ Ⓒ Ⓓ Ⓔ

67. Ⓐ Ⓑ Ⓒ Ⓓ Ⓔ
68. Ⓐ Ⓑ Ⓒ Ⓓ Ⓔ
69. Ⓐ Ⓑ Ⓒ Ⓓ Ⓔ
70. Ⓐ Ⓑ Ⓒ Ⓓ Ⓔ
71. Ⓐ Ⓑ Ⓒ Ⓓ Ⓔ
72. Ⓐ Ⓑ Ⓒ Ⓓ Ⓔ
73. Ⓐ Ⓑ Ⓒ Ⓓ Ⓔ
74. Ⓐ Ⓑ Ⓒ Ⓓ Ⓔ
75. Ⓐ Ⓑ Ⓒ Ⓓ Ⓔ
76. Ⓐ Ⓑ Ⓒ Ⓓ Ⓔ
77. Ⓐ Ⓑ Ⓒ Ⓓ Ⓔ
78. Ⓐ Ⓑ Ⓒ Ⓓ Ⓔ
79. Ⓐ Ⓑ Ⓒ Ⓓ Ⓔ
80. Ⓐ Ⓑ Ⓒ Ⓓ Ⓔ

81. Ⓐ Ⓑ Ⓒ Ⓓ Ⓔ
82. Ⓐ Ⓑ Ⓒ Ⓓ Ⓔ
83. Ⓐ Ⓑ Ⓒ Ⓓ Ⓔ
84. Ⓐ Ⓑ Ⓒ Ⓓ Ⓔ
85. Ⓐ Ⓑ Ⓒ Ⓓ Ⓔ
86. Ⓐ Ⓑ Ⓒ Ⓓ Ⓔ
87. Ⓐ Ⓑ Ⓒ Ⓓ Ⓔ
88. Ⓐ Ⓑ Ⓒ Ⓓ Ⓔ
89. Ⓐ Ⓑ Ⓒ Ⓓ Ⓔ
90. Ⓐ Ⓑ Ⓒ Ⓓ Ⓔ
91. Ⓐ Ⓑ Ⓒ Ⓓ Ⓔ
92. Ⓐ Ⓑ Ⓒ Ⓓ Ⓔ
93. Ⓐ Ⓑ Ⓒ Ⓓ Ⓔ
94. Ⓐ Ⓑ Ⓒ Ⓓ Ⓔ

95. Ⓐ Ⓑ Ⓒ Ⓓ Ⓔ
96. Ⓐ Ⓑ Ⓒ Ⓓ Ⓔ
97. Ⓐ Ⓑ Ⓒ Ⓓ Ⓔ
98. Ⓐ Ⓑ Ⓒ Ⓓ Ⓔ
99. Ⓐ Ⓑ Ⓒ Ⓓ Ⓔ
100. Ⓐ Ⓑ Ⓒ Ⓓ Ⓔ
101. Ⓐ Ⓑ Ⓒ Ⓓ Ⓔ
102. Ⓐ Ⓑ Ⓒ Ⓓ Ⓔ
103. Ⓐ Ⓑ Ⓒ Ⓓ Ⓔ
104. Ⓐ Ⓑ Ⓒ Ⓓ Ⓔ
105. Ⓐ Ⓑ Ⓒ Ⓓ Ⓔ

PRACTICE TEST 2

1. Soil forms as a result of which of the following processes?

   (A) Erosion of weathered material

   (B) Movement of material due to gravity

   (C) Movement of worms and burrowing mammals in the soil

   (D) Weathering of bedrock

   (E) Compaction and cementation of sediment

2. Which of the following was the geologic era when the dinosaurs roamed Earth?

   (A) Jurassic

   (B) Paleozoic

   (C) Cenozoic

   (D) Mesozoic

   (E) Cretaceous

3. Phosphorus is a primary ingredient in which of the following?

   (A) Pesticides

   (B) Herbicides

   (C) Fertilizers

   (D) Gasoline additives

   (E) Genetically modified crops

4. Which of the following is a significant phosphorus sink?

   (A) Deep ocean sediments

   (B) Nodules on legumes

   (C) Vegetation, especially boreal forests

   (D) Ocean water

   (E) The atmosphere

5. High levels of which nutrient can lead to eutrophication?

   (A) Carbon

   (B) Nitrogen

   (C) Oxygen

   (D) Carbon dioxide

   (E) Water

6. What role do consumers play in the cycling of chemicals through an ecosystem?

   (A) Consumers incorporate chemicals from the nonliving environment into organic compounds.

   (B) Consumers break down dead organisms.

   (C) Consumers feed on producers and incorporate some of the chemicals into their bodies.

   (D) Consumers supply the soil, air, and water with inorganic chemicals.

   (E) Consumers use the chemicals from the ecosystem.

GO ON TO THE NEXT PAGE

7. Which of the following cycles is MOST influenced by nonliving processes?

(A) Phosphorus cycle
(B) Carbon cycle
(C) Oxygen cycle
(D) Nitrogen cycle
(E) Water cycle

8. Which of the following might be a trace element that is cycled through an ecosystem?

(A) Water
(B) Iron
(C) Hydrogen
(D) Nitrogen
(E) Phosphorus

9. Virtually all of the molecular oxygen in the atmosphere today came from which source?

(A) Volcanic eruptions
(B) Outgassing
(C) Primitive atmosphere
(D) Respiration of animal life
(E) Photosynthesis of blue-green bacteria, algae, and green plants

10. Which of the following is NOT a factor that influences temperatures on Earth?

(A) Latitude
(B) Altitude
(C) Cloud cover
(D) Distance from large bodies of water
(E) Ocean currents

11. What is the difference between weather and climate?

(A) Climate is the daily temperature and moisture conditions in a place, and weather is a long-term pattern.
(B) Weather occurs in the troposphere, and climate occurs in the entire atmosphere.
(C) Weather is the daily temperature and moisture conditions in a place, and climate is a long-term pattern.
(D) Climate occurs in the troposphere, and weather occurs in the entire atmosphere.
(E) The weather in an area determines the climate.

12. Why do modern giraffes have long necks?

(A) They stretch to reach the leaves high in the trees.
(B) Long-necked giraffes got more food and were therefore more likely to reproduce.
(C) Long-necked giraffes produce more offspring than shorter-necked giraffes.
(D) It happened by chance.
(E) As the trees grew, so did the animals' necks.

13. The human population on Earth was 2.5 billion in 1950 and 6.4 billion in 2004. This is an example of what kind of growth?

(A) Arithmetic
(B) Dramatic
(C) Geometric
(D) Logistic
(E) Exponential

GO ON TO THE NEXT PAGE

14. When does the exponential growth of a population occur?

   (A) When resources are limited
   (B) When species become physically crowded
   (C) When resources are not limited
   (D) When the growth rate decreases
   (E) When the population exceeds the carrying capacity

15. Which of the following terms describes an organism that has been locally depleted by human activity?

   (A) Vulnerable
   (B) Threatened
   (C) Endangered
   (D) Extinct
   (E) Rare

16. The phrase "survival of the fittest" is sometimes used to describe which of the following?

   (A) Evolution
   (B) Natural selection
   (C) Genetic variability
   (D) Mutations
   (E) Chance happenings in nature

17. Which of the following statements is TRUE about extinction?

   (A) Extinction is driven by human actions.
   (B) Extinction occurs in rapid bursts.
   (C) Extinction is a natural process.
   (D) There has been one major period of extinction on Earth.
   (E) If left alone, all extinctions in nature would cease.

18. Which of the following is a place where primary succession may occur?

   (A) New island created by volcanoes
   (B) Forest land cleared for farming
   (C) Area after forest fire
   (D) Dried up lake
   (E) Alaskan wilderness

19. What are the first organisms that may appear after a disturbance damages an existing community but leaves the soil intact?

   (A) Lichens
   (B) Autotrophic microorganisms
   (C) Shrubs
   (D) Grasses
   (E) Trees

20. Which is the broadest level of ecological study?

   (A) Ecosystem
   (B) Community
   (C) Biosphere
   (D) Population
   (E) Individual organism

21. Laws in the United States make it possible to dispose of and handle different types of solid waste. Which of the following is NOT an option under current U.S. laws?

   (A) Burying waste underground
   (B) Exporting waste to other countries
   (C) Burning waste
   (D) Dumping waste into the open ocean
   (E) Using landfills

GO ON TO THE NEXT PAGE

22. A body of water is considered to be safe for swimming if there are no more than 200 colonies of coliform bacteria per 100

    (A) liters of water.
    (B) microliters of water.
    (C) milliliters of water.
    (D) ppm of water.
    (E) gallons of water.

23. The dissolved oxygen content in water is usually measured in

    (A) parts per billion.
    (B) microliters.
    (C) degrees.
    (D) percentages.
    (E) parts per million.

24. Which of the following is the major source of solid waste in the United States?

    (A) Farming
    (B) Domestic products
    (C) Mining waste
    (D) Industry
    (E) Municipal systems

25. How long does it typically take an aluminum beverage can to decompose?

    (A) 1 to 2 weeks
    (B) 10 to 30 days
    (C) 1 year
    (D) 100 to 500 years
    (E) 1 million years

26. Which of the following contributes to the formation of ozone?

    (A) Sulfur dioxide
    (B) Nitric oxide
    (C) Nitrous oxide
    (D) Carbon dioxide
    (E) Carbon monoxide

27. How are point sources of pollution different from nonpoint sources?

    (A) Point sources are harder to remediate.
    (B) It is difficult to place liability for point sources.
    (C) Point sources are concentrated.
    (D) Nonpoint sources are easy to remediate.
    (E) Nonpoint sources are generally localized.

28. At which of the following ranges in pH will the ecology of a lake environment begin to be significantly impacted?

    (A) 3.0–3.5
    (B) 4.5–5.0
    (C) 5.0–6.0
    (D) 6.5–9.0
    (E) 9.0–9.5

29. The NIMBY defense can become an issue when people consider disposing of solid waste in which of the following ways?

    (A) Composting
    (B) Demanufacturing
    (C) Detoxifying
    (D) Landfills
    (E) Ocean dumping

GO ON TO THE NEXT PAGE ▷

30. The majority of the municipal solid waste generated in the United States today is

    (A) recycled.
    (B) recovered and reused.
    (C) placed in landfills.
    (D) burned.
    (E) dumped in the ocean.

31. Biological agents can be transmitted to humans in a variety of ways. Which of the following routes of transmission have the highest risk?

    (A) Ingestion
    (B) Aerosol routes
    (C) Occupational exposure
    (D) Incidental exposure
    (E) Skin contact

32. What was the first anti-pollution law passed in the United States?

    (A) Migratory Bird Treaty Act
    (B) Rivers and Harbors Act
    (C) Oil Pollution Act
    (D) Clean Water Act
    (E) Clean Air Act

33. The Environmental Protection Agency says that an acceptable risk to the public from carcinogenic food additives, such as pesticides, is

    (A) 1 in 5,000.
    (B) 1 in 100,000.
    (C) 1 in 1 million.
    (D) 1 in 10 million.
    (E) 1 in 1 billion.

34. What is risk characterization?

    (A) A way to evaluate risk according to personal choice
    (B) An estimate of damage from varied and chronic doses
    (C) An assessment of what damage may occur and where damage will be done
    (D) A way to determine how much of a chemical is absorbed
    (E) The use of data from different methods to determine a hazard risk

35. Air pollution returning to the surface of Earth may return as

    (A) a solid only.
    (B) a liquid only.
    (C) a gas only.
    (D) a solid, liquid, or gas.
    (E) acid rain only.

36. The eggs and small fry of some aquatic organisms are damaged when a rapidly melting snow pack contains acidic particles, resulting in acidic concentrations of 5 to 10 times higher than acidic rainfall. This condition is known as

    (A) acid shock.
    (B) eutrophication.
    (C) leaching.
    (D) ozone depletion.
    (E) synergistic effects.

37. Acid rain may leach nutrients from the soil. This could lead to

    (A) an increase in calcium content in soil.

    (B) a decrease in plant growth.

    (C) an increase in lead and cadmium in the plants.

    (D) problems in absorbing water.

    (E) more susceptibility to insect damage.

38. What effect can acid rain have on an aquatic ecosystem?

    (A) Leaching of Al^{3+} ions

    (B) Destruction of metals and stone

    (C) Increase in animal life

    (D) Decline in plant growth

    (E) Nitrogen saturation

39. High levels in surface water of which of the following compounds increases the risk of acid shock to aquatic life?

    (A) Iron

    (B) Ozone

    (C) Lead

    (D) Aluminum

    (E) Oxygen

40. Acid precipitation leads to an increase in which two elements in the soil?

    (A) Sulfur and oxygen

    (B) Magnesium and lead

    (C) Aluminum and magnesium

    (D) Sulfur and nitrogen

    (E) Carbon and oxygen

41. Peroxyacyl nitrates are produced from the reaction of hydrocarbons, oxygen, and nitrogen oxides. These are examples of

    (A) point source pollutants.

    (B) nonpoint source pollutants.

    (C) primary pollutants.

    (D) secondary pollutants.

    (E) inorganic pollutants.

42. At the mouth of the Mississippi River, there is an area that is depleted in oxygen. What is the most likely cause of this?

    (A) Acid precipitation falling in the ocean

    (B) Reduction in plant life

    (C) Fertilizers being carried by the river

    (D) Thermal pollution

    (E) Overfishing

43. The process by which a body of water becomes enriched in dissolved nutrients, stimulating the growth of aquatic plant life and usually resulting in the depletion of dissolved oxygen, is called

    (A) acid shock.

    (B) acid deposition.

    (C) algal blooms.

    (D) eutrophication.

    (E) ultraviolet oxidation.

44. What impact do algal blooms have on plants living deeper beneath the surface of the water?

    (A) They have no impact.

    (B) They supply extra nutrients to deeper-growing plants.

    (C) They kill them.

    (D) They increase the amount of sunlight to deeper-growing plants.

    (E) They cause deeper-growing plants to seek other forms of food.

45. Which of the following pH levels is harmless to most fish?

    (A) 3.0

    (B) 4.0

    (C) 5.0

    (D) 6.0

    (E) 8.0

GO ON TO THE NEXT PAGE

46. Which of the following reactions becomes slower as altitude increases?

   (A) $O_2 + hv \rightarrow O + O$
   (B) $O + O_2 \rightarrow O_3$
   (C) $O_3 + hv \rightarrow O_2 + O$
   (D) $O + O_3 \rightarrow O_2 + O_2$
   (E) $O_2 + O_2 \rightarrow O + O_3$

47. An increase in evaporation at lower latitudes could lead to

   (A) a decrease in precipitation.
   (B) lower temperatures.
   (C) a decrease in the number of hurricanes.
   (D) an increase in the number of hurricanes.
   (E) unpredictable hurricane seasons.

48. What is the most common form of skin cancer?

   (A) Melanoma skin cancer
   (B) Squamous skin cancer
   (C) Basal skin cancer
   (D) Sarcoma
   (E) Lymphoma

49. The lenses of your eyes absorb UVB radiation. An increase in UVB radiation has been shown to have a link to

   (A) nearsightedness.
   (B) farsightedness.
   (C) cataracts.
   (D) retina problems.
   (E) increased pupil dilation.

50. Over the past 60 years, the upper 300 meters of ocean water has warmed an average of

   (A) 0.06°C.
   (B) 0.3°C.
   (C) 1.0°C.
   (D) 3.0°C.
   (E) 6.0°C.

51. Which ocean has the warmest surface temperatures?

   (A) Pacific
   (B) Arctic
   (C) Indian
   (D) North Atlantic
   (E) South Atlantic

52. What develops in the oceans between warm surface waters and cooler bottom waters?

   (A) Thermocline
   (B) Heat gradient
   (C) Isocline
   (D) Contour
   (E) Heat differential

53. In which direction do easterlies push surface water near the equator?

   (A) Eastward
   (B) Northward
   (C) Southward
   (D) Westward
   (E) Down

GO ON TO THE NEXT PAGE

54. Which of the following is TRUE about subsurface currents?

(A) They travel faster than surface currents.

(B) They are not affected by water densities.

(C) They travel slower than surface currents.

(D) They are unaffected by temperature differences.

(E) They are unaffected by differences in salinity.

55. Which of the following areas would be MOST affected by a rise in sea level?

(A) Wetlands

(B) Forests

(C) Grasslands

(D) Deserts

(E) Tundra

56. Models have suggested that temperatures will rise faster and higher in which of the following regions?

(A) Equatorial regions

(B) Midlatitudes

(C) Deserts

(D) Polar regions

(E) Rain forests

57. Global warming has caused El Niños to

(A) last for shorter periods of time.

(B) occur more frequently.

(C) impact a more narrow area.

(D) produce colder ocean temperatures.

(E) produce less flooding.

58. Which aspect of volcanic eruptions has the MOST impact on climate?

(A) They release carbon dioxide.

(B) They introduce lots of sulfur dioxide into the atmosphere.

(C) They release gas and ash.

(D) The hot lava is released as thermal pollution.

(E) They change the landscape.

59. The majority of ocean pollution comes from which of the following sources?

(A) Runoff from pesticides, fertilizer, and sewer discharges

(B) Acid rain

(C) Marine sources such as oil spills

(D) Ocean dumping

(E) Oil exploration and production

60. Which of the following sources of CH_4 is being reduced?

(A) Burning of biomass

(B) Cattle and other animals

(C) Wetlands

(D) Anthropogenic activity

(E) Extraction of natural gas and oil

61. Species that are transported to areas where they do not naturally live are called

(A) endangered species.

(B) threatened species.

(C) native species.

(D) indigenous species.

(E) invasive species.

GO ON TO THE NEXT PAGE

62. Which of the following is the largest contributor to declining fish catches?

    (A) Global warming
    (B) Rising sea level
    (C) Lower reproductive rates
    (D) Pollution
    (E) Overfishing

63. Which of the following is the BEST term for the range of genetic differences found within and between species?

    (A) Biodiversity
    (B) Genetic drift
    (C) Ecological diversity
    (D) Genetic diversity
    (E) Biotic relationships

64. Where is diversity at its greatest?

    (A) At the equator
    (B) At the midlatitudes
    (C) At the poles
    (D) In the middle of continents
    (E) On the edges of continents

65. Which of the following would lead to an increase in biodiversity?

    (A) Agriculture and logging
    (B) Introduction of exotic species
    (C) Decreasing geographic isolation
    (D) Increases in temperature
    (E) Increased capture rates for fishing

66. Using pesticides and practices such as monocultures

    (A) increases biodiversity.
    (B) decreases biodiversity.
    (C) has no impact on biodiversity.
    (D) increases productivity.
    (E) decreases productivity.

67. The SI unit of energy is the

    (A) British thermal unit.
    (B) kilocalorie.
    (C) Joule.
    (D) calorie.
    (E) Calorie.

68. Which of the following fishing methods has the lowest bycatch?

    (A) Pole and line fishing
    (B) Long-line fishing
    (C) Trawling
    (D) Drift netting
    (E) Gill netting

69. A man picks an apple from an apple tree that has been growing in the sun and eats it. In this example, which of the following is the energy source?

    (A) Man
    (B) Apple
    (C) Tree
    (D) Earth
    (E) Sun

70. A thermometer measures what type of energy?

    (A) Nuclear
    (B) Kinetic
    (C) Potential
    (D) Electrical
    (E) Heat

GO ON TO THE NEXT PAGE ⇨

71. Which of the following choices BEST completes this statement to make it true: The quality of energy _____ as it flows through an ecosystem because of the _____ .

    (A) increases; first law of thermodynamics
    (B) decreases; first law of thermodynamics
    (C) does not change; first law of thermodynamics
    (D) decreases; second law of thermodynamics
    (E) increases; second law of thermodynamics

72. Which of the following devices would be used for aquaculture?

    (A) Purse-seine
    (B) Cage
    (C) Trawl bag
    (D) Drift net
    (E) Lines

73. Which of the following methods of increasing the fish harvest could have direct adverse consequences on the food web?

    (A) Fish farming
    (B) Harvesting squid, octopus, krill, and unconventional species
    (C) Cutting waste
    (D) Fish ranching
    (E) Increasing shellfish harvests

74. An even disappearance of topsoil is a sign of which type of erosion?

    (A) Wind erosion
    (B) Gully erosion
    (C) Sheet erosion
    (D) Rill erosion
    (E) Subsidence erosion

75. Which of the following soil conservation methods involves planting different crops in alternating rows to trap soil and catch runoff?

    (A) Conventional-tillage farming
    (B) Strip cropping
    (C) Conservation-tillage farming
    (D) Contour farming
    (E) Terracing

76. Which of the following soil conservation methods would be appropriate for dry arid regions?

    (A) Terrace farming
    (B) Strip cropping
    (C) Windbreaks
    (D) Contour farming
    (E) Crop rotation

77. Which of the following is a problem with captive breeding programs?

    (A) All animals breed well in captivity.
    (B) It is easy to move animals from one zoo to another for breeding.
    (C) It is difficult to determine the sex of many animal species.
    (D) Captive breeding programs are inexpensive.
    (E) Few assisted reproductive technologies are available.

GO ON TO THE NEXT PAGE

78. What is an objection to the growth and use of genetically engineered (GE) fruits and vegetables?

   (A) GE fruits and vegetables taste different from normal ones.

   (B) GE fruits and vegetables will only grow in the laboratory, not on large-scale farms.

   (C) GE fruits and vegetables will attract more pests than normal ones.

   (D) GE fruits and vegetables can pass genes on to native crops when grown in the wild and upset the ecological balance.

   (E) GE fruits and vegetables have no market value.

79. The Green Revolution made extensive use of which of the following techniques?

   (A) Genetically engineered crops

   (B) Integrated pest management

   (C) Chemicals (fertilizers, pesticides)

   (D) Crop rotation

   (E) Intercropping

80. Which of the following is NOT a feature of sustainable agriculture?

   (A) Extensive use of fertilizers and pesticides

   (B) Planting multiple crops together

   (C) Harvesting crops at different times

   (D) Maintaining the natural environment

   (E) Never leaving the plot bare

81. Which of the following statements is TRUE?

   (A) Most of the world's population relies heavily on meat for their diets.

   (B) In the United States, most of the crops produced are for human consumption.

   (C) In LDCs, grains are the most important food staple.

   (D) Our bodies make all of the vitamins and minerals that we need to survive.

   (E) According to the United Nations, the number of undernourished people in the world is decreasing.

82. A fuel cell would be an essential part of which of the following alternative energy sources?

   (A) Solar power

   (B) Hydrogen power

   (C) Geothermal power

   (D) Wind power

   (E) Hydroelectric power

83. Which of the following is NOT a feature of a passive solar-heated house?

   (A) Building part of the structure into the ground (Earth-sheltering)

   (B) Storing heat in water-filled columns or barrels

   (C) Roof-mounted passive heaters

   (D) Facing the house north in the Northern Hemisphere

   (E) Attaching a greenhouse to the structure

84. Which of the following methods is used MOST commonly in North America to harvest trees?

   (A) Slash and burn

   (B) Clear-cutting

   (C) Selective cutting

   (D) Shelterwood cutting

   (E) Seed-tree cutting

GO ON TO THE NEXT PAGE

85. Which of the following is NOT a good method to protect forests from natural diseases and pathogens?

(A) Clear-cutting infected areas

(B) Maintaining biodiversity

(C) Importing timber trees to replace infected ones

(D) Developing disease-resistant trees

(E) Integrated pest management

86. What is the greatest danger facing national parks today?

(A) Resource conservation

(B) Increasing predators (wolves, bears, etc.)

(C) Alien species

(D) Human activities

(E) Resource management

87. The U.S. Census Bureau has the following information on the populations of five countries for last year. The data are shown below.

Which country has the greatest population growth for last year?

(A) Country A

(B) Country B

(C) Country C

(D) Country D

(E) Country E

88. Examine the age-structure diagrams and choose the best statement.

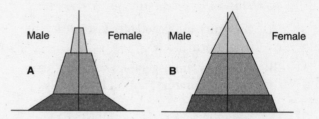

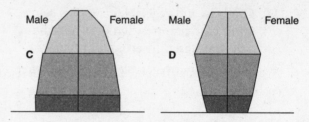

(A) Country A probably has an effective antinatalist policy.

(B) Country B has an even male-to-female ratio at all ages.

(C) Country C is an example of a rapidly growing population.

(D) Country B is an example of negative population growth.

(E) Country D probably has an effective antinatalist policy.

	Birth Rate (No./1,000/yr.)	Death Rate (No./1,000/yr.)	Immigration Rate (No./1,000/yr.)	Emigration Rate (No./1,000/yr.)
Country A	20	20	10	5
Country B	10	5	20	10
Country C	30	20	5	15
Country D	65	50	5	10
Country E	10	5	10	10

GO ON TO THE NEXT PAGE

89. The United States has a slowly growing population. What is the major factor that accounts for its population growth?

(A) Rapidly increasing birth rates

(B) Rapidly increasing death rates

(C) Rapidly increasing immigration rates

(D) Rapidly increasing emigration rates

(E) Rapidly decreasing death rates

90. Examine the demographic transition graph below and identify the statement below that is TRUE.

(A) In the Transitional phase, both birth and death rates are high and the population grows rapidly.

(B) In the Pre-Industrial phase, death rates drop, but birth rates remain high.

(C) In the Industrial phase, death rates increase and approach the birth rates.

(D) In the Post-Industrial phase, there is negative population growth.

(E) In the Industrial phase, there is negative population growth.

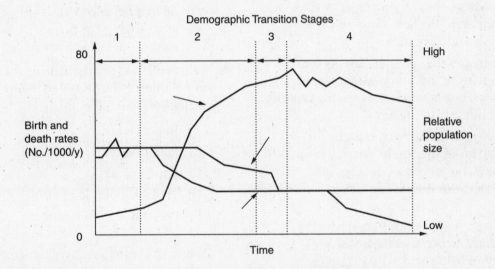

91. Which of the following is MOST true about a demographic trap?

(A) A country remains in Stage 1.

(B) A country cannot transition from Stage 3 to Stage 4.

(C) A country remains in Stage 3.

(D) A country falls back from Stage 4 to Stage 3.

(E) A country cannot transition from Stage 2 to Stage 3.

92. What country has the highest population?

(A) United States

(B) France

(C) South Africa

(D) China

(E) Mexico

GO ON TO THE NEXT PAGE

93. Which of the following information is NOT part of demographics?

(A) Age
(B) Birth rate
(C) Sex
(D) Geographic region
(E) Economic status

94. Damming the Colorado River in the United States has had which of the following effects?

(A) No effect on the carrying capacity for humans of the Southwestern United States and Mexico
(B) Increases the carrying capacity for humans of the Southwestern United States and Mexico
(C) Decreases the carrying capacity for humans of the Southwestern United States and Mexico
(D) Decreases the carrying capacity for humans of the Southwestern United States, but increases the carrying capacity for humans in Mexico
(E) Increases the carrying capacity for humans of the Southwestern United States, but decreases the carrying capacity for humans in Mexico

95. Which pair of words BEST completes the following sentence: Countries with a _____ standard of living tend to have _____ populations.

(A) low, decreasing
(B) low, stable
(C) high, high
(D) high, low
(E) high, increasing

96. Which of the following is a non-use benefit of a coastal environment?

(A) Whale watching
(B) Oyster harvesting
(C) Fishing
(D) Wildlife observation
(E) Eel grass preservation zone

97. In the early years of the Industrial Revolution, soot settling on trees in England helped dark colored moths avoid predators, causing the population to shift from lighter to darker colored moths. This is an example of

(A) disruptive selection.
(B) reproductive isolation.
(C) stabilizing selection.
(D) directional selection.
(E) ecological isolation.

98. Which method of internalizing the cost of pollution is the MOST economically inefficient?

(A) Impact fee
(B) Pollution tax
(C) Emission tax
(D) Tradable pollution permit
(E) Direct pollution limits

99. Which of the following regulations was the first legislation to address protecting cultural resources?

(A) Surface Mining Control and Reclamation Act
(B) Antiquities Act
(C) Clean Air Act
(D) Endangered Species Act
(E) National Historic Sites Act

GO ON TO THE NEXT PAGE

100. Which of the following is a characteristic associated with the Endangered Species Act of 1973?

   (A) It seeks to preserve all species.

   (B) Species on the list can be collected but not hunted or killed or injured.

   (C) All commercial shipments of wildlife and wildlife products can enter or leave through any port.

   (D) Federal agencies may not fund, authorize, or carry out projects that jeopardize, threaten, modify, or destroy habitats of a species on the endangered species list.

   (E) Decisions to list or to unlist a species must be made on the basis of economics and biology.

101. Which of the following laws requires a social impact statement to assess cultural resources as well as an environmental impact statement for any federal activity?

   (A) Clean Air Act

   (B) Endangered Species Act

   (C) National Environmental Policy Act

   (D) Comprehensive Environmental Response, Compensation, and Liability Act

   (E) Surface Mining Control and Reclamation Act

102. Which Federal Agency researches the health effects of pesticides and regulates their use?

   (A) National Institute of Environmental Health Sciences

   (B) Department of Agriculture

   (C) Bureau of Land Management

   (D) Environmental Protection Agency

   (E) Occupational Safety and Health Administration

103. Which of the following is NOT a provision of the Law of the Sea?

   (A) Established an Exclusive Economic Zone of 12 miles for coastal nations

   (B) Covers rights to fishing, marine life, and mining

   (C) Establishes general guidelines for safeguarding marine life and protecting the freedom of scientific research on the high seas

   (D) Creates a legal mechanism for controlling mineral resource exploitation in international waters

   (E) Created the International Seabed Authority

104. Local climate, water, soil, geology, and biology play an important role in which principle of sustainable land-use strategy?

   (A) Time principle

   (B) Disturbance principle

   (C) Place principle

   (D) Species principle

   (E) Land cover principle

105. A wildlife corridor would be an important part of which of the following?

   (A) Sustainable land-use strategies

   (B) Bioremediation

   (C) Conservation easement

   (D) Debt for Nature swap

   (E) Land reclamation

IF YOU FINISH BEFORE TIME IS CALLED, YOU MAY CHECK YOUR WORK ON THIS SECTION ONLY. DO NOT TURN TO ANY OTHER SECTION IN THE TEST.   STOP

FREE-RESPONSE QUESTIONS

1. A class wanted to study the effect of radiation on seed germination. They grew radish seeds provided by a biological supply company. The seeds had been exposed to three levels of radiation from a Cobalt-60 source. The plants were grown in a greenhouse under stable conditions over a one-week period. For each data set and for each level of exposure, 25 seeds were planted. As each plant was counted, it was marked so that it was not counted a second time. The students collected the information below.

Radish Seeds Germinated					
	Exposure (Krads)	Number of seeds germinated			
		26 hrs.	80 hrs.	100 hrs.	168 hrs.
data set 1	control	2	0	5	8
	50,000	0	2	5	6
	150,000	0	0	2	7
	500,000	0	0	0	2
data set 2	control	2	6	9	9
	50,000	1	4	5	7
	150,000	0	1	8	6
	500,000	0	0	0	4
data set 3	control	3	8	9	9
	50,000	0	6	8	9
	150,000	0	7	7	8
	500,000	0	0	1	5
data set 4	control	3	4	4	4
	50,000	1	3	4	5
	150,000	0	3	5	5
	500,000	0	0	1	1

(A) (3 points) What is the significance of the control? Identify the constants that should be included as part of the experimental setup.

(B) (4 points) Present the data as percentages germinated for each radiation level over time. Show mathematical calculations.

(C) (2 points) Using the graph below, graph a compilation of the data showing radiation level vs. total plants germinated by the end of the experiment.

(D) (2 points) Draw a conclusion statement about this experiment that relates your data to the world outside of the classroom. Cite one case in which radiation became an environmental problem.

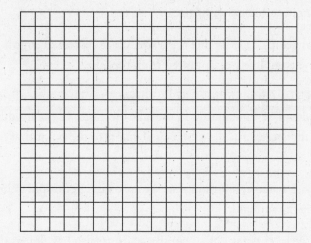

2. **Farming**

The Green Revolution of the 1940s increased food production all over the world by creating a system of high-input agriculture that requires increased amounts of pesticides and fertilizers. These in turn have created problems in the soil, air, and water, and for wildlife. Future increases in food production employing the techniques of the Green Revolution are limited. Today, genetic modifications are being sought to continue to increase the production of food.

(A) (3 points) Pesticide use can lead to ecological imbalances. Explain what happens and how this can be prevented.

(B) (3 points) Define sustainable agriculture and list two practices that make current agricultural methods more sustainable.

(C) (3 points) Genetic modification may be fantastically valuable to some crops. Identify two applications of genetic modification to agriculture. Why should we be concerned about GM (genetically modified) crops, animals, or products?

(D) (3 points) Discuss two ways that retain and/or preserve the genetic diversity of plants and animals used in our food.

3. Ecology/Restoration

Cichlids are a fish that provided a major food source for the people who live around Lake Victoria, which borders Uganda, Tanzania, and Kenya in Africa. There were numerous varieties occupying all the niches of the lake. Many varieties were extremely colorful.

In the 1960s, the Nile perch was introduced to the lake in an effort to provide another food source for the fishers of the lake. The perch multiplied very successfully, but the devastating consequences of the introduction of this alien species on the cichlid populations were not recognized for about two decades because of the political turmoil in eastern Africa.

The Nile perch has a voracious appetite due to its size, up to six feet in length. It feeds on cichlids, so a number of varieties have completely disappeared from the lake.

Preparing the Nile perch for consumption involves drying, which requires firewood from the surrounding forest. As trees are cut down, sedimentation in the lake increases. In the muddy waters, the cichlids have evolved with a significant decrease in the vibrancy of colors to the point that many of these varieties have become brown.

An explosion of algae and water hyacinth in the lake are causing eutrophication of the lake, choking the waterways and blocking the sunlight.

Development around the lake, typical of all populated areas, also has altered the lake environment.

(A) (3 points) Describe two niches in a lake that the cichlids may have occupied.

(B) (2 points) Explain the evolutionary mechanism by which turbidity in a lake would have an effect on the colors of fish in a lake.

(C) (2 points) Restoration of the lake may be impossible, but some positive steps can be taken to improve water quality in the lake. At this point, what can be done to improve the ecosystem of Lake Victoria?

(D) (3 points) Explain how the bloom in algae and water hyacinth affects the cichlid populations.

4. Incineration is a common method of waste disposal. The degree to which incinerators create unhealthful environmental conditions for people living in close proximity to them varies greatly depending on the pollution prevention equipment employed in building them.

(A) (3 points) Label the diagram of the mass burn incinerator shown in the picture.

(B) (2 points) Describe one pollution control device that is part of a mass burn incinerator.

(C) (3 points) Manufacturers have made a concerted effort to use less material in the production of certain products. Describe one example.

(D) (2 points) Name and elaborate on one piece of legislation that addresses (non-nuclear) waste disposal.

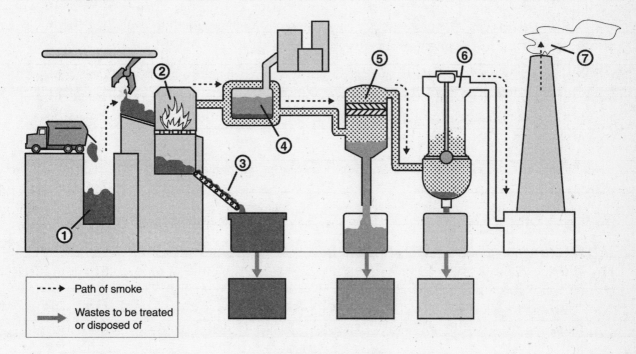

Mass Burn Incinerator

_____

_____

_____

_____

_____

_____

_____

_____

_____

_____

_____

_____

_____

_____

_____

_____

_____

_____

_____

_____

_____

IF YOU FINISH BEFORE TIME IS CALLED, YOU MAY CHECK YOUR WORK ON
THIS SECTION ONLY. DO NOT TURN TO ANY OTHER SECTION IN THE TEST.

STOP

ANSWERS AND EXPLANATIONS

1. D

Choice (E) can be eliminated because that describes the formation of sedimentary rocks. Choices (A) and (B) discuss weathering and erosion that, while part of some soil formation, do not describe how soil is formed. Choice (C) seems to be a likely candidate, but notice how the soil is already formed. Worms and other animals within the soil will mix it up and aerate it, not form it.

2. D

Dinosaurs existed during the Jurassic and Cretaceous, but these are geologic periods, not eras. The Mesozoic was the era that contained these periods.

3. C

Phosphorous can be used to help plants grow, making it a good fertilizer.

4. D

Unlike the carbon and nitrogen cycles, no pool of phosphorus exists in the atmosphere, so the phosphorus cycle is limited mainly to the soil and water. This eliminates nodules on legumes, (B), and vegetation, (C), as well as the atmosphere, (E). Aquatic plants and phytoplankton assimilate phosphate. Marine vertebrates and invertebrates eat plants, phytoplankton, and each other. Bacterial decomposers break down organic phosphate to inorganic phosphate, which returns to the water.

5. B

Eutrophication is the rapid growth of algae in bodies of water usually due to high levels of either nitrogen or phosphorus.

6. C

A producer, not a consumer, would incorporate chemicals, (A and E), from the environment into organic compounds. Decomposers break down dead organisms, (B), and supply the soil, air, and water with inorganic chemicals, (D).

7. E

The phosphorous, carbon, oxygen, and nitrogen cycles all depend heavily on living things. A small portion of the water cycle (transpiration) does depend on plants, but this dependence is very limited.

8. B

The other choices—water, (A), hydrogen, (C), nitrogen, (D), and phosporus, (E)—are major elements that cycle through the environment and a specific ecosystem.

9. E

Volcanic eruptions, (A), and outgassing, (B), describe the sources of much of the water vapor in the atmosphere. The primitive atmosphere, (C), consisted mainly of hydrogen and helium. Respiration, (D), would increase the amount of carbon dioxide in the atmosphere.

10. C

Latitude, (A), and altitude, (B), refer to the relationship between Earth and the Sun, which is directly responsible for the changing of the seasons. The temperatures of large bodies of water, (D), and ocean currents, (E), also affect global temperatures. While cloud cover may locally change temperatures for a brief period of time, cloud cover does not cause significant change in temperatures.

11. C

Weather is a short-term phenomenon (such as rain, wind, or sunny conditions) in a localized area, while climate is the long-term weather patterns.

12. B

The long necks of today's giraffes formed over millions of years through natural selection, eliminating choices (A), (D), and (E), which suggest a more immediate reason. In times of drought, the slightly longer-necked individuals could reach higher vegetation and survive, but the short-necked ones would die off. In the next generation, there would be variations again, but the average neck length might be slightly longer than the average neck length of the previous generation. So while offspring is an important factor in natural selection, it is not sheer number of offspring, (C), but the number of offspring that survive because of adaptations to the environment that matter.

13. E

This rapid and significant increase in population is an example of exponential growth.

14. C

If resources are limited, (A), or a species becomes crowded, (B), the growth rate will decrease, not increase. The same will happen if the growth rate decreases, (D), or if a population exceeds the carrying capacity, (E).

15. A

A threatened species, (B), may become endangered. An endangered animal, (C), is close to becoming extinct. An extinct species, (D), no longer exists on Earth. And a rare species, (E), is just not found in many places, but is not in danger.

16. B

Survival of the fittest is another way of describing natural selection. It means that the animal that has adapted best to the environment through mutations, (D), will procreate. It has been "selected by nature" to survive. Evolution, (A), is change over time. Genetic variability, (C), is simply various genotypes found in a species, due to the mutations being passed

on to new generations. Chance happenings in nature are random mutations that are not a purposeful adaptation, (E).

17. C

Extinction occurs naturally on Earth, and it occurred millions of years before the arrival of humanity, eliminating choices (A) and (E). There has been more than one "period of extinction" or mass extinction on Earth (the most famous is that of the dinosaurs), and many are gradual, background extinctions of species, eliminating (B) and (D).

18. A

Primary succession occurs in a virtually lifeless area with no soil. All of these areas, except for choice (A), already have soil and some sort of life.

19. D

Autotrophic microorganisms, (B), would be the first thing to appear in an area with no life and no soil. Lichens, (A), would appear as part of primary succession. The area mentioned here is undergoing secondary succession. Shrubs, (C), and trees, (E), would come after the grasses.

20. C

The biosphere is the sum of all Earth's ecosystems. An ecosystem, (A), includes abiotic and biotic factors in an area. A community, (B), is all the organisms living in a particular area. A population, (D), is a group of individual organisms, (E), of the same species in a particular area.

21. D

Burying waste in a sanitary landfill is an option used by many communities, so choice (A) is incorrect. Open landfills, (E), while not as popular as they once were, are still used. Waste is sometimes exported, (B), although this can be prohibitive. Incineration, (C), is also used, which leaves choice (D) as the only possible answer.

22. C

Colonies of coliform bacteria are safe if there are no more than 200 colonies per 100 milliliters. Microliters, (B), are too small a measurement, and gallons, (E), and liters, (A), are too large. Parts per million of water, (D), is not an accurate way to express amounts of water.

23. E

Degrees, (C), and percentages, (D), are used to measure different things. Microliter, (B), and parts per billion, (A), while used to measure substances within another, are not correct.

24. C

More solid waste is created in the United States by mining practices than by any other method.

25. D

Organic waste (vegetable and fruit peels) takes about one to two weeks, (A), to decompose. Paper goods take an average of 10 to 30 days, (B), to decompose. Items made of wool take up to 1 year, (C), and it is estimated that plastic bags can take up to 1 million years. It takes tin, aluminum, and other metal items about 100 to 500 years to decompose.

26. B

Nitric oxide is a major player in the upper atmosphere in the formation of ozone (O_3). Sulfur dioxide, (A), nitrous oxide, (C), carbon dioxide, (D), and carbon monoxide, (E), do not contribute significantly to the formation of ozone.

27. C

The remediation of a pollution problem is not a factor in the definition of whether it is a point or a nonpoint source, eliminating choices (A) and (D). A point source of pollution is a single, identifiable *localized* source of air, water, thermal, noise, or light pollution, while nonpoint sources come from many unidentifiable sources, eliminating choices (B) and (E).

28. C

High pH levels would not show much of a change in the ecology of a lake environment, choices (D) and (E), which are close to or above a neutral pH reading of 7.0. The detrimental effects of lowering levels of pH in an aquatic ecosystem become pronounced at around a pH of 6.0. Choices (A) and (B) are well below this level, so the impact would already be significantly showing.

29. D

NIMBY stands for "Not in my backyard," in this case meaning that while people want to change the way that solid waste is disposed of, they don't want it disposed of near their homes. Composting, (A), involves setting up a "compost pile" where organic material that degrades quickly is deposited. Many people use this small-scale disposal method in their own homes. Demanufacturing, (B), and detoxifying, (C), take place in industrial areas, and the consequences of ocean dumping, (E), would not directly affect the aesthetics or real estate value of a neighborhood. These would not be NIMBY issues for homeowners.

30. C

It is illegal to dump waste into the ocean in this country, eliminating choice (E). Of all the municipal solid waste generated in the United States, 57 percent is placed in landfills, 28 percent is recycled, (A), or recovered and reused, (B), and 15 percent is burned, (D).

31. B

Biological agents can be transmitted in all these ways. The spread of these agents through aerosols, choice (B), is the most significant. Some people may receive their exposure at work, choice (C), or by the use of things like cosmetics, choice (D). They may get it from food or water, choice (A), or by skin contact, choice (E), but from contact in the air is the most significant.

32. B

The Rivers and Harbors Act, (B), passed in 1899, banned disposal of refuse into water bodies and is widely considered to be the first environmental protection law in the United States. You can easily eliminate choice (C) even if you don't know that the Oil Pollution Act was passed in 1989, simply because oil spills are a relatively recent problem. Choice (A) is a very early environmental law, but it regulated hunting and commerce, not pollution. Choices (D) and (E) were both passed in the late 1960s to early 1970s, long after the Rivers and Harbors Act.

33. C

The Delaney Clause of the EPA's Federal Food, Drug and Cosmetic Act Food Additives Amendment of 1958 calls for zero risk to the public from carcinogenic food additives. However, because zero risk cannot be achieved at a reasonable cost, the EPA usually has permitted carcinogenic food additives if no more than one in a million individuals consuming such food additives over their lifetimes would be expected to die of cancer resulting from exposure to these chemicals.

34. E

Risk characterization includes evaluating risk according to personal choice (A), estimating damage, (B), assessing damage, (C), and an in-depth look at chemical interactions, (D). Because risk characterization looks at all of these factors, choice (E) is correct.

35. D

Acid deposition can be in the form of solids, liquids (including acid rain), and gases, so choice (D) is correct. For example, there can be acid precipitation in the form of snow or rain, and there can also be acid fog.

36. A

Acid shock occurs when acids are released into the water over a short period of time at concentrations that are more acidic than rainfall. Adult fish can survive this shock, but the eggs and small fry are vulnerable. Eutrophication is the growth of algae, making choice (B) incorrect. Leaching, (C), refers to the movement of materials out of another material. Ozone depletion, (D), is not directly tied to this problem, and synergistic effects, (E), is too broad an answer.

37. B

Because leaching is the removal of material, it would not increase the calcium content in the soil, eliminating choice (A). An increase in lead and cadmium, (C), problems absorbing water, (D), and insect damage, (E), while serious effects of pollution on plants, are not the result of leaching.

38. D

Leaching of aluminum ions occurs in soil, eliminating choice (A). Metals and stone are impacted by acid rain, but not in aquatic ecosystems, eliminating choice (B). Acid rain decreases the growth and survival of aquatic animals and plants, eliminating choice (C). Acid rain does not cause nitrogen saturation, eliminating choice (E) and leaving choice (D) as the correct answer.

39. D

Both the lower pH and higher aluminum concentrations in surface water occur as a result of acid rain. This can cause damage to fish and other aquatic animals.

40. D

Acid rain does not add oxygen to the soil, eliminating choices (A) and (E). It also does not add magnesium, eliminating choices (B) and (C). Only choice (D) contains two elements that in fact come from acid rain, sulfur and nitrogen.

41. D

There is no mention of the precise source of these materials, which eliminates choices (A) and (B). Hydrocarbons, oxygen, and nitrogen oxides are organic materials, eliminating choice (E). They are also primary pollutants, making choice (C) a tempting choice. However, the question is asking about the peroxyacyl nitrates that are the product of the reaction of these materials. Secondary pollutants form when primary pollutants react in the atmosphere, making choice (D) the correct answer.

42. C

The consequences of acid rain falling on the ocean, (A), would not show up in that way. There is no indication of thermal pollution, (D), or overfishing, (E), in the mouth of the Mississippi. A reduction in plant life, (B), is tempting, but it would more likely be a result of less oxygen, not a cause. The Mississippi River carries fertilizers and pesticides from the surrounding farmlands and dumps them into the Gulf of Mexico, so (C) is the correct choice. These chemicals create an area that is depleted of oxygen.

43. D

Eutrophication is the enrichment in dissolved nutrients. Acid shock is what happens to aquatic plants and animals when more acid materials are quickly added to the ecosystem. Acid deposition, (B), is the depositing of acid materials. Algal blooms, (C), are the result of eutrophication. Ultraviolet oxidaton, (E), is the result of extended exposure to sunlight.

44. C

The algae does not supply deeper-growing plants with food or allow sunlight to reach them, eliminating choices (B) and (D). Therefore, algal blooms do have an impact on deeper plants, which eliminates choice (A). The plants have no other possible food source, which makes choice (E) incorrect. This ultimately will kill the deep water plants, (C).

45. E

The ecology of an aquatic ecosystem begins to be deleteriously altered by a pH of 6.0, eliminating choice (D). Fish eggs, fry, and carp are adversely affected by a pH of 5.0, (C). Trout, whitefish, salmon, and smelt are killed in an environment with a pH of 4.0, (B). A pH of 3.0, (A), is toxic to most fish.

46. B

Ozone is the result of a reaction between elemental oxygen and oxygen gas, choice (B). Choice (A) is an example of a Chapman reaction. Choice (C) becomes faster with altitude. Choice (D) is a formula that describes the destruction of ozone. Choice (E) is not a formula for the production of ozone at all.

47. D

As evaporation increases in the lower latitudes, temperatures will rise, eliminating choice (B). This rise in temperature would ultimately result in an increase in precipitation in some areas, not a decrease, eliminating choice (A). There is no indication about the impact of the predictability of hurricanes, (E), but the increase in the number of them is likely, eliminating choice (C).

48. C

Sarcoma, (D), is a cancer of the connective tissues, and lymphoma, (E), is a cancer that grows in the lymph nodes. Melanoma skin cancer, (A), and squamous skin cancer, (B), are less common than basal skin cancer, (C).

49. C

UVB radiation has no direct impact on your eyesight, eliminating nearsightedness, (A), and farsightedness, (B). Increased exposure to UVB radiation can lead to problems with the lens of your eyes. This means that choice (D) about retinal problems is incorrect. While exposure to sunlight may temporarily impact the dilation of your pupils, (E), this is not a consequence of UVB radiation. Cataracts, (C), are problems with the eye's lens that can lead to blindness.

50. B

The upper 300 meters of the oceans have warmed an average of about 0.3°C over the past 60 years, choice (B). Water in the upper 3,000 (not 300) meters has warmed an average of 0.06°C, (A), over the same period. Temperature increases of 1.0°C, 3.0°C, and 6.0°C (choices (C), (D), and (E), respectively) are too high.

51. C

The Indian Ocean, (C), has the highest surface temperatures because most of its area is in the tropics. The tropics receive more sunlight than any other region on Earth, so the surface waters of the Indian Ocean are heated more than those of any other ocean.

52. A

A thermocline, (A), is a line that separates different temperatures. An isocline, (C), is a bend in a series of rocks. Contour lines, (D), connect points of equal elevation on Earth's surface. Heat gradients, (B), and heat differentials, (E), describe ranges of temperatures and are not "produced" when conflicting temperatures meet.

53. D

Easterlies are the prevailing winds that blow from the high-pressure areas of the North and South Poles toward low-pressure areas at around 60° latitude. That air is then deflected eastward by the Coriolis Effect, pushing the waters westward.

54. C

Subsurface currents are more susceptible to temperature differences and differences in salinity than surface currents, eliminating choices (D) and (E) as correct answers. The subsurface currents move slower than surface currents and are affected more by density, pointing to choice (C) as the correct answer and eliminating choices (A) and (B).

55. A

Regions of forests, (B), grasslands, (C), desert, (D), and tundra, (E), tend to be more inland. Because wetlands, (A), are often found along the coast, they are much more affected by changes in sea level.

56. D

Because of the arrangement of Earth with respect to the incoming solar radiation as well as current conditions at the poles, temperature differences will be greatest there.

57. B

In general, El Niños occur every 10 to 20 years. But recently, El Niños have occurred more frequently, have been more dramatic, and have lasted longer than normal, eliminating choices (A) and (E). The effects of El Niños can reach as far as Africa, eliminating choice (C). The El Niños produce higher ocean temperatures, eliminating choice (D). For example, the event of 1982–1983 produced ocean temperatures of up to 11°F above normal.

58. B

The discharge of lava into ocean areas, (D), and the changing of the landscape, (E), have significantly more impact on Earth's surface rather than climate. The release of carbon dioxide, (A), as well as gas and ash, (C), can have a serious impact on climate, but the introduction of sulfur dioxide into the atmosphere is, by far, the most influential.

59. A

Of all ocean pollution, 44 percent comes from land-based discharges, such as runoff. Acid rain accounts for 33 percent of the pollution in the ocean. Marine sources such as oil spills account for 12 percent of ocean pollution. Ocean dumping is responsible for 10 percent, and oil exploration accounts for 1 percent of the pollution in the ocean.

60. C

Methane production from wetlands is going down because the practice of wetland destruction is decreasing. All other sources of methane are increasing, including burning of biomass, (A), cattle, (B), human activity, (D), and extraction of natural gas and oil, (E).

61. E

Native species, (C), and indigenous species, (D), are organisms that naturally live in an area. Endangered species, (A), and threatened species, (B), are descriptions of how many organisms of a certain species still survive. Invasive, (E), or exotic, species are introduced to an area.

62. E

All of the choices except rising sea level, (B), play a role in declining fish catches, and even sea level rise has the potential to disrupt fisheries by harming nursery habitats. However, of all these options, overfishing, (E), has the largest impact on fisheries' decline.

63. D

Genetic drift, (B), is a change in the gene pool of a population due to chance. Ecological diversity, (C), refers to the variety of ecosystems and ecological communities. Biotic relationships, (E), are the different parts of the living environment. Biodiversity, (A), is more a measure of the variety among organisms. The best answer is genetic diversity, as it deals specifically with genetic differences within species.

64. A

Diversity is greatest in tropical regions found near the equator, (A), and decreases toward the poles, (B and C). There is no indication that biodiversity is greater or less at the shore, (E), compared to the middle of a continent, (D).

65. C

Agriculture and logging, (A), introducing exotic species, (B), temperature increases, (D), and increased capture rates, (E), can lead to a decrease in biodiversity, as they are disruptive and deleterious events to an environment.

66. B

Pesticides and monocultures decrease biodiversity. The impact of pesticides and monocultures on productivity is unclear.

67. A

A Joule, (C), is the basic SI unit of energy (0.24 cal). The British thermal unit, (A), is an English system measurement, the amount of energy necessary to raise the temperature of 1 pound of water by 1°F. A calorie, (D), is a metric unit that is the amount of energy required to raise the temperature of 1 gram of water by 1°C. A kilocalorie, (B), and a calorie, (E), are the same thing—1,000 calories.

68. A

All non-target fish and other organisms, such as turtles and whales, are referred to as bycatch. Choices (B), (C), (D), and (E) are all non-selective fishing methods that can catch any fish in the target area, making it difficult to limit the catch exclusively to target species. Of the choices, only pole and line fishing, (A), is a selective method: it gives fishers the ability to choose which fish they are catching and limits bycatch.

69. E

The man is the energy sink, so choice (A) is incorrect. The apple and tree are intermediate steps in the chain, so choices (B) and (C) are incorrect. Earth is not involved directly in the chain, so choice (D) is incorrect. In this example, solar energy, (E), is captured by the tree through photosynthesis and is stored as sugars in the apple. The man eats the apple and utilizes the energy.

70. B

A thermometer measures the kinetic energy of the molecules in the object in contact with the thermometer. Heat energy is the energy transferred from one object to another because of a difference in temperature, so, while close, choice (E) is incorrect.

71. D

The first law of thermodynamics deals with the conservation of energy and matter, not the quality of energy; therefore, choices (A) through (C) are not correct. As energy flows through an ecosystem, the quality of energy decreases with each transformation (trophic level) according to the second law of thermodynamics; therefore, choice (E) is incorrect and choice (D) is correct.

72. B

Purse-seines, (A), drift nets, (D), and lines, (E), are commercial fishing methods used for catching pelagic fish. Trawl bag, (C), is a commercial fishing method for catching bottom dwelling fish. Cages and fences, (B), are used for aquaculture methods of fish farming and fish ranching.

73. B

Fish farming, (A), and ranching, (D), are types of aquaculture that create artificial systems where fish breed and grow, but do not affect the food web. Increasing shellfish harvests, (E), does not increase the fish catch. Reducing waste, (C), would increase the fish harvest from those already caught and would not alter the food web. Increased harvesting of squid, octopus, and krill could potentially deplete the lower trophic levels of the food web and reduce the harvests of fish, so choice (B) is correct.

74. C

Soil erosion by wind, (A), would not be an even disappearance of topsoil. Subsidence does not cause erosion, so choice (E) is incorrect. Water erosion is classified into gully, rill, and sheet erosions. Of these, gully and rill form from rapidly flowing water that cuts channels into the soil and erodes it unevenly, so choices (B) and (D) are incorrect. Only sheet erosion causes gradual uniform loss of topsoil, so choice (C) is correct.

75. B

Conventional farming, (A), and conservation-tillage farming, (C), deal with plowing the land and exposing or reducing the exposure of topsoil. Contour plowing, (D), and terracing, (E), deal with how the plowing of the land follows the natural slopes of the land to reduce runoff. Only strip cropping involves planting alternate rows of a soil-conserving ground cover crop, such as grasses or grass/legume mixes with the desired agricultural crop (corn, wheat, etc.).

76. C

In dry, arid regions, wind is the major source of erosion, not running water. Terrace farming, strip cropping, and contour farming are all methods of conserving soil from water runoff, so choices (A), (B), and (D) are incorrect. Crop rotation is an important method of maintaining soil fertility, so choice (E) is incorrect. Windbreaks, (C), appropriately reduce the velocity of wind movements across farmlands in dry areas, thereby reducing soil erosion.

77. C

Not all animals breed well in captivity, so choice (A) is incorrect. Captive breeding programs are expensive to maintain, so choice (D) is incorrect. In many instances, animals must be transported from one zoo to another for breeding; this process is expensive and poses risks to the animals, making choice (B) incorrect. New assisted reproductive technologies are available that have helped the success of many captive breeding programs, so choice (E) is incorrect. In many animals, especially birds and reptiles, it is difficult to tell males and females apart without surgical examination or biochemical tests. This difficulty has hampered many captive breeding programs.

78. D

GE fruits and vegetables do not taste different and do have market value, so choices (A) and (E) are incorrect. GE crops can grow in the wild, so choice (B) is incorrect. GE crops do not necessarily attract more pests; in fact, some are engineered with natural pesticide genes, so choice (C) is incorrect. Many fear that GE crops will pass genes on to native plants in the wild, thereby causing new, unwanted genetic variants and upsetting the natural ecological balance, making (D) the correct choice.

79. C

The Green Revolution occurred before the advances in genetic engineering and integrated pest management, so choices (A) and (B) are incorrect. Crop rotation and intercropping are methods for soil conservation and not integral to the Green Revolution, so choices (D) and (E) are incorrect.

80. A

Sustainable agriculture applies ecological principles to farming. Multiple crops are planted at the same time, (B), which means that they are harvested at different times, (C), and the plots are never bare, (E), leaving the environment in a natural state, (D). Sustainable agriculture relies on natural pesticides and biodiversity for pest control, not fertilizers and pesticides, so choice (A) is correct.

81. C

In the United States, most agricultural products are produced for livestock, not humans, so choice (B) is incorrect. Our bodies do not make all of the vitamins and minerals necessary for us to survive, so choice (D) is incorrect. According to the United Nations, the number of undernourished people in the world is increasing, not decreasing, making choice (E) incorrect. Most of the world's population relies on grains for their diet, (C), and not meats, so choice (A) is incorrect.

82. B

A fuel cell is a major aspect of the hydrogen economy. Hydrogen and oxygen react in a membrane to produce electricity and water (a by-product).

83. D

Passive solar heating designs optimize heat collection from the Sun and convert it into low temperature heat for space heating. Earth-sheltering provides insulation and heat storage, so choice (A) is incorrect. Passive solar designs utilize heat storage containers such as water-filled containers and roof-mounted heat collectors, so choices (B) and (C) are incorrect. Attaching a greenhouse to the structure aids in solar collection and heat distribution, so choice (E) is incorrect. Building a home in the Northern Hemisphere that faces north reduces the ability to collect sunlight, so choice (D) is correct.

84. B

Slash and burn is a method used to clear the land for agriculture, but it does not harvest the trees, making choice (A) incorrect. In selective cutting, (C), small clusters of mature trees are cut down. Shelterwood cutting, (D), takes mature trees in several cuttings over many years. Both these methods are expensive and not commonly used. Seed-tree cutting cuts nearly all of the trees, but it leaves only a few seed-producing trees uniformly distributed. It is a good ecological way to harvest trees, but not the most commonly used, making choice (E) incorrect. Most of North America's forests are cleared by cutting and harvesting all of the trees in an area—clear-cutting, (B).

85. C

Clear-cutting diseased areas and burning the trees helps reduce the infested areas, so choice (A) is incorrect. Maintaining natural biodiversity and using integrated pest management techniques are the best ways to reduce pests and pathogens, so choices (B) and (E) are incorrect. Developing disease-resistant trees is helpful in reducing pathogens, so choice (D) is incorrect. Importing timber trees to replace diseased ones increases the risk of introducing new pathogens into a forest and would not be a good method of reducing or preventing diseases and pests in forests, so choice (C) is correct.

86. D

The greatest danger to parks today is human activities (mining, logging, grazing, pollution, etc.), so choice (D) is correct. While resource conservation and management activities are ongoing concerns of park rangers, they are not the most important problem facing parks, so choices (A) and (E) are incorrect. The number of predators in parks is actually decreasing because of hunting, so choice (B) is incorrect. Alien species pose a problem to parks but not the most important one, so choice (C) is incorrect.

87. B

This question assesses your ability to use the equation for change in human population growth: $\Delta N_t = (B + I) - (D + E)$, when given data on rates of birth, death, immigration, and emigration. When you use the equation, Countries A and E have net population increases of 5 people/1,000/yr., so choices (A) and (E) are incorrect. Country D shows a net population increases of 10 people/1,000/yr., so choice (D) is incorrect. Country C has no increase in population. In fact, its net population change is zero, so choice (C) is incorrect. Country B shows a net increase of 15 people/1,000/yr., which is the greatest increase, so choice (B) is correct.

88. E

Country A clearly has a rapidly growing population with high birth rates, so it does not have an effective antinatalist (anti-birth) policy, so choice (A) is incorrect. Country B has an uneven male-to-female ratio and is an example of a slowly growing population, so choices (B) and (D) are incorrect. Country C is an example of a zero growth population, so choice (C) is incorrect. Country D shows a negative growth rate with birth rates less than death rates; it is a likely candidate for having an effective antinatalist policy, so choice (E) is correct.

89. C

The United States has a slowly increasing growth rate. Rapidly increasing birth rates and rapidly decreasing death rates theoretically would contribute to population growth, but are not characteristic of the United States, so choices (A) and (E) are incorrect. Rapidly increasing death and emigration rates would contribute to a population decrease, not increase, so choices (B) and (D) are incorrect. The major factor contributing to the increase in the U.S. population growth is increasing immigration, so choice (C) is correct.

90. D

The Pre-Industrial phase (1) has high birth rates, high death rates, and low population growth rate, so choice (B) is incorrect. In the Transitional phase (2), death rates drop, birth rates remain high, and the population grows rapidly, so choice (A) is incorrect. In the Industrial phase (3), birth rates drop and approach the death rates, thereby causing population growth to slow, so choices (C) and (E) are incorrect. In the Post-Industrial phase (4), birth rates decline to at or below the death rates (negative population growth) and the population size decreases, so choice (D) is correct.

91. E

In some cases, rapid growth rates and deterioration of the environment reduce the standard of living and impoverish the population. Poverty brings a sense of helplessness to the citizens, and they rely on large families (more children) to help their economic situation. These conditions prevent the country from progressing out of Stage 2 and have been referred to as a *demographic trap.*

92. D

The five most populous countries in the world in order from number 1 to number 5 are China, India, the United States, Indonesia, and Brazil, so China, (D), is correct.

93. B

Demography is the field that describes the statistics of populations. Demographic data from censuses and data are broken down into vital statistics (age groups, sex, economic status, geographic regions, etc.). Birth rates are data on population dynamics, not demography, so choice (B) is correct.

94. E

Humans live in geopolitical borders, not environmental ones. Natural resources, such as rivers, forests, and lakes, can cross national boundaries. So, by utilizing common resources, one country can affect another country's population (e.g., damming rivers, burning forests, etc.). The Southwestern United States and Mexico share the Colorado River. Damming the river in the United States would have beneficial effects on the Southwestern United States and detrimental effects on Mexico, so choices (A) through (D) are incorrect. Such a project would make more water available to the Southwestern United States and less to Mexico, resulting in an increased carrying capacity of the Southwestern United States and decreasing that of Mexico. Choice (E) is correct.

95. D

Countries with low standards of living (third-world countries or LDCs) have rapidly increasing populations, so choices (A) and (B) are incorrect. Countries with a high standard of living utilize more resources and have lower populations, so choices (C) and (E) are incorrect, while choice (D) is correct.

96. E

Nonuse benefits are those we get from knowing that an environment has been preserved. The eel grass preservation zone, (E), is an example of a nonuse benefit. A direct use benefit is one that provides a service or product to humankind. The use benefits can be direct (products) or indirect (recreation). Coastal ecosystems can provide products in the forms of fish, (C), and oysters, (B), or recreation, such as whale watching, (A), and wildlife observations, (D).

97. D

Choices (B) and (E) can both be ruled out since they refer to isolating mechanisms that can separate species from each other, which is not applicable here. Choice (A) refers to natural selection that selects for both extremes over an average; with the moths described in the question, this would mean that both light *and* dark moths increased over gray moths, which did not occur. Stabilizing selection, (C), selects for the average over the extremes; with the moths described, this would mean that gray moths increased over light and dark moths, which again did not occur. That leaves choice (D), directional selection, where an environmental change gives an advantage to a particular extreme, which is clearly the case in this example. Note that with recent improvements in air quality, the moths have begun shifting back toward lighter variants.

98. E

An externality is something that does not affect the producer of a good but influences society as a whole. A negative externality diminishes the standard of living of the society. Pollution is an example of a negative externality because the polluting company loses no money, but society must pay to clean up the pollution. To internalize the costs, the polluters must pay the costs of polluting through impact fees, taxes, and tradable permits issued by the government. These methods are economically efficient because the company must compare the costs of pollution (e.g., cleanup) versus the revenues of the product to determine if it is profitable to continue polluting, so choices (A) through (D) are incorrect. Pollution limits force companies to clean up in conformance with the government standards without regard to marginal costs and marginal revenues, which is very inefficient for the company, so choice (E) is correct.

99. B

The Surface Mining Control and Reclamation Act, the Clean Air Act, and the Endangered Species Act all address environmental concerns, so choices (A), (C), and (D) are incorrect. Both the Antiquities Act

and the National Historic Sites Act address cultural resources; however, the Antiquities Act, (B), was passed in 1906 and the National Historic Sites Act, (E), was passed in 1935.

100. D

The Endangered Species Act of 1973 was the first legislation to give legal standing to wildlife. Because the act does not seek to preserve all wildlife, only those species on the endangered list, choice (A) is incorrect. The act provided that species on the list could not be hunted, killed, injured, *or* collected, making choice (B) incorrect. The commercial shipments of wildlife and wildlife products were regulated through nine designated U.S. ports, making choice (C) incorrect. Decisions as to whether a species is listed or unlisted as endangered must be made based solely on biology, not economics, so choice (E) is incorrect. According to the act, federal agencies could not fund, authorize, or conduct projects that would damage habitats of endangered species, so choice (D) is correct.

101. C

The National Environmental Policy Act stressed the government's obligation toward preservation, both environmental and cultural; this act required all government agencies to submit both environmental and social impact statements for any proposed federal activity, so choice (C) is correct. The Clean Air Act made states within the United States set air quality standards and implement plans to curb pollution, so choice (A) is incorrect. The Endangered Species Act authorized government agencies (Fish and Wildlife Service, National Marine Fisheries Service) to identify and list endangered and threatened species, both terrestrial and aquatic, so choice (B) is incorrect. The Comprehensive Environmental Response, Compensation, and Liability Act authorized the federal government to clean up hazardous waste sites and respond to accidental/disastrous releases of hazardous wastes into the environment, so choice (D) is incorrect. The Surface Mining Control and Reclamation Act of 1977 requires mining companies to replant vegetation on land that was strip-mined, so choice (E) is incorrect.

102. D

The Department of Agriculture, the Bureau of Land Management, and the Occupational Safety and Health Administration do not research or regulate pesticide use, so choices (B), (C), and (E) are incorrect. While the National Institute of Environmental Health Sciences, (A), may research the health effects of pesticides, only the Environmental Protection Agency, (D), researches and regulates pesticide use.

103. A

The Law of the Sea was ratified by 60 nations and is in effect. It governs rights to fishing, marine life, and minerals on the seafloor within a coastal nation's Exclusive Economic Zone (EEZ), so choice (B) is incorrect. It provides for protection of marine life and scientific research on the high seas, so choice (C) is incorrect. It also establishes a regime for mineral resources in international waters that is enforced by The International Seabed Authority, making choices (D) and (E) incorrect. The EEZ for coastal nations was extended from 3 miles to 200 miles, so choice (A) is correct. The territorial waters of a coastal nation extend for 12 miles, not its EEZ.

104. C

The place principle involves local climate, water, soil, geology, and biologic interactions and how they strongly affect both the functioning of the ecosystem and the abundance/distribution of species in any one place, so choice (C) is correct. The time principle recognizes that various ecological processes occur over different time scales, so choice (A) is incorrect. The disturbance principle states that characteristics of the disturbance—type, intensity, duration—shape the characteristics of the ecosystem, so choice (B) is incorrect. The species principle recognizes that the interactions of species and certain species within an ecosystem have important broad effects on that ecosystem, so choice (D) is incorrect. The land-cover principle states that the land cover—size, shape, spatial relationship—influences the populations, communities, and ecosystems, so choice (E) is incorrect.

105. A

A wildlife corridor is a portion of developed land that retains its natural habitat for biodiversity and is a part of a sustainable land-use strategy, (A). Bioremediation and land reclamation are ways of fixing the land after use, so choices (B) and (E) are incorrect. Conservation easements and Debt for Nature swaps are ways of preserving the land without use or development, so choices (C) and (D) are incorrect.

ANSWERS TO FREE-RESPONSE QUESTIONS

1. For this question, the answers are very straightforward. On the surface, it may seem too easy, and the student will try to make more of it than it is. This answer is purely about experimental method and could be answered by any science student.

    (A) A control acts as a basis for comparison. It is the setup that is kept at normal conditions. In this case, it is the group of seeds that was not irradiated.

    Constants are the factors that are the same for every seed, such as temperature of the greenhouse, amount of water given to each seed, frequency of the watering, light, and size of container with soil or growing medium.

    (B) This is straight addition and division. There are four sets of seeds for each radiation level, each with 25 seeds, equaling 100. Therefore, adding up the number of seeds that germinated actually yields the percentage. The problem is simple, but it tests an ability to manipulate numerical data.

Compilation Data					
Exposure (Krads)	Number of Seeds Germinated				% Germinated
	26 hrs.	80 hrs.	100 hrs.	168 hrs.	
Control	10	18	27	30	85
50,000	2	15	22	27	66
150,000	0	11	22	26	59
500,000	0	0	2	12	14

    (C) Check to see that your scale is consistent and your axes labeled properly, with the radiation levels on the x-axis (independent variable) and the % germination on the y-axis (dependent variable). Each of the four data points must be plotted clearly.

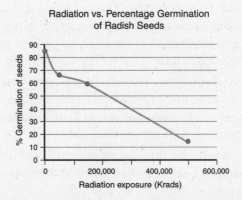

Radiation vs. Percentage Germination of Radish Seeds

    (D) The obvious conclusion is that a large amount of radiation exposure is bad for living things. The nuclear accident at Chernobyl in the former Soviet Union is the case that most students will consider. The scare after the accident at Three Mile Island in Pennsylvania is the other case commonly cited in textbooks.

2. (A) Pesticides can lead to imbalances by creating resistant strains of pests. One of the four mechanisms of evolution is that genetic diversity exists in each generation. A resistant individual will survive the pesticide and go on to further produce other resistant individuals. This leads to a resistant population that is no longer affected by the pesticide. Other imbalances are created with broad-spectrum pesticides, which may kill off not only the pest but also the natural predator of a pest, thereby creating a new problem where none existed before the pesticide was used. Two other methods to reduce pesticides are the use of natural predators (such as ladybugs) instead of pesticides, and using the "scout and spray" method of pesticide application rather than simply spraying on a calendar schedule regardless of need.

(B) Sustainable agriculture consists of practices that can yield food indefinitely without depleting the resources needed in production. Two good practices in sustainable agriculture are (1) the use of good organic manure, which releases nutrients more slowly, and (2) intercropping plants of more than one type so that pests cannot spread unrestrained.

(C) Genetically modified foods may in fact yield fantastic results, such as by embedding pesticides within a plant to combat a disease that would destroy the crop. The reason that we should be concerned is the unintended and unforeseen ecological effects. Plant resistance may also develop to these plants.

(D) Seed banks and planting native varieties for commercial use are two methods that preserve genetic diversity and plants and animals. Allowing areas to remain undeveloped, although very obvious, is also a viable answer.

3. This question is document based, so at least some of the answers are embedded in the text. Always look for the clues.

(A) Cichlid niches would include multiple feeding methods, such as bottom feeding. Some occupy rocky bottoms and others live on sandy bottoms. The multicolored varieties would likely have lived closer to the sunlit waters at the surface, as bottom feeders are usually the same color as the sediment.

(B) The conditions needed for evolution are overproduction, variation, limits of growth (struggle for survival), and differential success for reproduction. When fish are brightly colored, that may be an adaptation for attracting mates. When the water is not clear and the colors are not visible, those with the brightest colors may have no advantage for mating, so there is no selection for this trait. If less brightly colored individuals are as likely to pass on genes as those that are dull colored, then the population evolves into a less vividly colored one.

(C) Planting trees around the lake, massive removal of the water hyacinth, and increased fishing of the Nile perch would be positive steps toward restoration of the lake. Limits on development around the lake and proper sewage disposal would help. Constant monitoring and good funding from governments for restoration efforts are necessary.

(D) As the algae and water hyacinth die, they use up the dissolved oxygen in their decomposition, depleting the dissolved oxygen needed by fish. Many of the fish will most likely die.

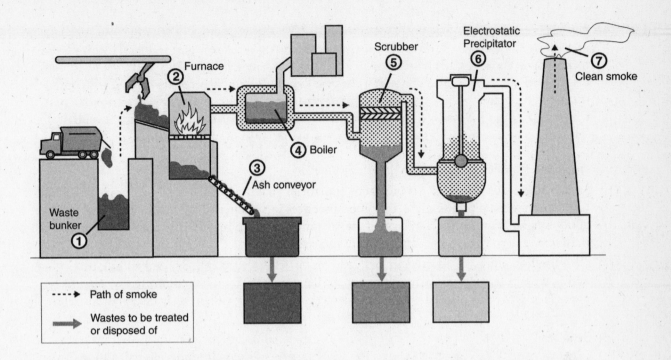

4. (A) and (B) The parts of the mass burn incinerator are (1) the waste bunker where the waste is deposited, (2) the furnace where it is burned, (3) an ash conveyor on which the ash is collected for deposition in a landfill, and (4) a boiler that creates steam used for electricity or heat. (5) Pollution control devices are the scrubbers in which a spray of lime neutralizes acidic gases, and (6) electrostatic precipitators, which give the ash a positive charge so it can stick to negatively charged plates. (7) Finally, smoke is released via the smoke stack.

   In part A, some credit would be awarded for partial labeling.

   (C) The reduction of waste at its origin is an underused strategy used by manufacturers. Dematerialization, the use of less material for a given product as the technology improves, is another idea that reduces the quantity of waste. Slimmer packaging, such as that used on CDs, smaller computers, thinner television sets, and MP3 players with downloaded music instead of numerous CDs are all examples.

   (D) The Comprehensive Environmental Response, Compensation and Liability Act (CERCLA), commonly called the Superfund, requires the parties responsible for creating a hazardous waste site be responsible for its cleanup. Unfortunately, toxic waste sites are often identified long after they were created and the responsible parties are long gone, so the government is left with the bill for cleanup.

   Some other pieces of legislation include the Resource Conservation and Recovery Act of 1976 (RCRA), which deals with municipal and industrial nonhazardous waste; the Marine Plastic Pollution Research and Control Act of 1987, which focuses on dumping of garbage in waters close to coastlines of the United States; and the Pollution Prevention Act of 1990, which was the first U.S. law that focused on source reduction.